AF440704

GRAPHICAL TECHNIQUES FOR
MULTIVARIATE DATA

Graphical Techniques for Multivariate Data

B. S. Everitt

NORTH-HOLLAND

NEW YORK

NORTH-HOLLAND

NEW YORK

© B. S. Everitt 1978

Library of Congress Cataloging in Publication Data
Everitt, B. S.
   Graphical techniques for multivariate data.
Includes index.
1. Multidimensional scaling – graphic methods.
2. Psychological research.   3. Medical research.
4. Social science research.   1. Title
(DNLM: 1. Computers, digital.   2. Factor analysis,
statistical.   3. Data display.   4. Sociometric
technics. BF39 E93G)
BF39.E93   150'.7'23   78–3108

ISBN 0 444 19461 4

Printed in Great Britain

Contents

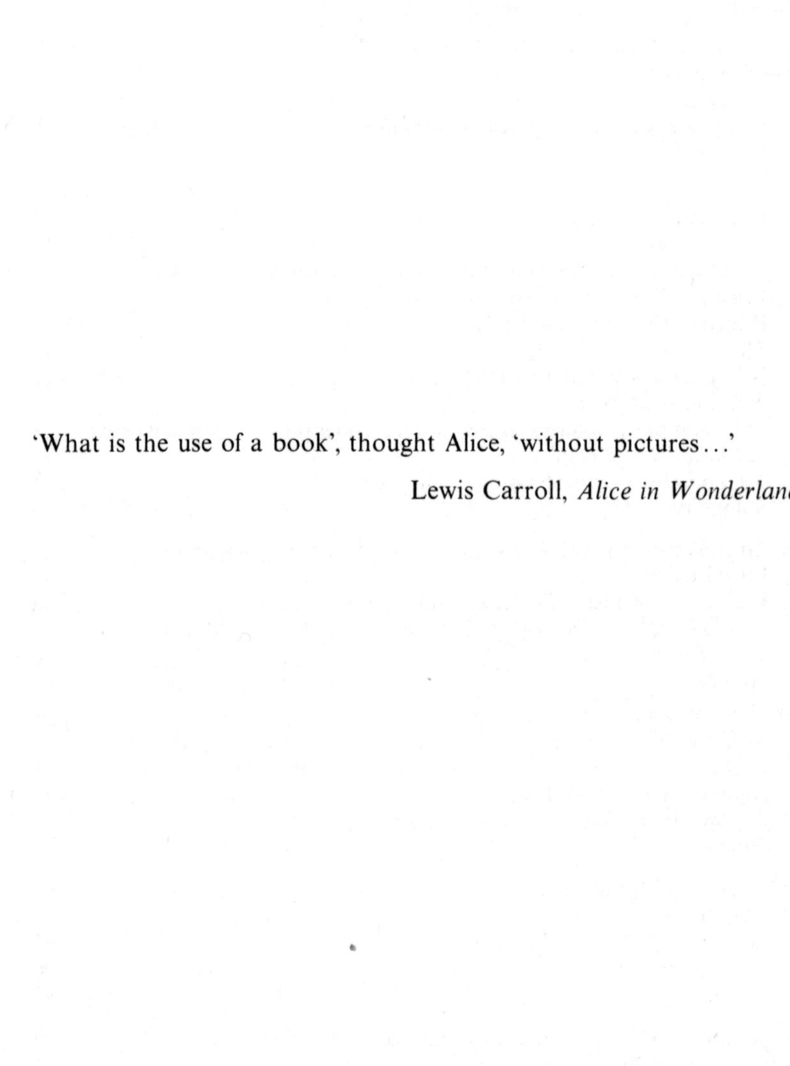

'What is the use of a book', thought Alice, 'without pictures...'

Lewis Carroll, *Alice in Wonderland*

Preface

This book is intended to serve as an introduction to some techniques for examining multivariate data which have in common that they produce their results in the form of a graph, diagram, or some other type of visual output. For none of the techniques to be described is it *essential* to have access to any specialized type of computer graphics hardware, although in many cases a plotting device of relatively high precision, for example, a microfilm plotter, is a very convenient tool for producing the diagrams, etc.

The text is aimed at research workers in psychology, psychiatry, medicine, and the social sciences, and others dealing with the complexities of multivariate data. For this reason the mathematical sections are kept at a fairly low level, although *some* knowledge of elementary statistics and matrix algebra is assumed.

My thanks are due to Nortei Omaboe who helped with some of the programming, to Professor A. E. Maxwell who read the manuscript and made many helpful comments, especially concerning those aspects of the English language that seem destined to remain forever a mystery to the author, and to Mrs Bertha Lakey who typed the manuscript.

B. S. Everitt

Institute of Psychiatry
London, 1977

1. Graphical Techniques and Multivariate Data

Introduction

'The preliminary examination of most data is facilitated by the use of diagrams. Diagrams prove nothing, but bring outstanding features readily to the eye; they are therefore no substitute for such critical tests as may be applied to data, but are valuable in suggesting such tests, and in explaining conclusions founded upon them.'

R. A. Fisher, *Statistical Methods for Research Workers*

The graphical examination of data has long been regarded by statisticians as an extremely useful technique and most would advocate it as an essential prerequisite to more formal analyses. Statistical diagrams serve several purposes; if we regard the three general objectives of data analysis as

1. Display and Description,
2. Analysis and Interpretation,
3. Summarization and Exposure,

then the corresponding graphical aids may be viewed as ranging from presentations of the raw data, through presentations of the results of fairly complex analyses, to graphical outputs which themselves constitute the statistical analysis by providing both a summary of the informational content of the data and an exposure of unanticipated characteristics, such as possible inadequacies of the assumed model. The use of diagrams may often reveal aspects of the data that might have escaped notice were the data tabulated only in numerical form; they may help to uncover features of the data that were totally unanticipated prior to the analysis, or to suggest hypotheses which may be further investigated.

Graphical techniques and statistical diagrams for use with *univariate data* are well known and are described briefly in many statistical textbooks, and in detail in Huff (1954). The main purpose of this text is to familiarize researchers in many areas with graphical techniques which may prove useful in investigating *multivariate data*. The term graphical is used to indicate any type of visual output rather than graphs *per se*.

1

Graphical Techniques and Multivariate Data

According to Gnanadesikan and Wilk (1969), data – by its very nature – is always multivariate since it involves observations associated with various facets of a particular background or environment. Certainly the data collected by today's researchers in areas such as psychology, psychiatry, and the social sciences are almost invariably multivariate. In general we represent the sample of multivariate data by a matrix $\mathbf{X}$ in which the ij-th element will represent the value of the j-th variable on the i-th individual. (The term 'individual' will be used in the text to denote the observational unit, whether this is actually human or otherwise.) That is

$$\mathbf{X} = \begin{bmatrix} x_{11} & \cdots & x_{1p} \\ x_{21} & & \\ \vdots & & \\ x_{n1} & \cdots & x_{np} \end{bmatrix}$$

where p is the number of variables and n is the number of individuals.

Many excellent texts are available which deal with the analysis of multivariate data using familiar methods such as factor analysis, discriminant function analysis, etc.; examples are those of Press (1972) and Kendall (1975). However, the formal numerical techniques of data analysis described in such texts are too often designed to yield specific answers to rigidly defined questions, and, in general, the objectives of data analysis are neither so narrow nor so formal as described and implied by some statistical theories of estimation and hypothesis testing. (This, of course, applies to all data but is perhaps *particularly* true for multivariate data which, because of the interrelated character of the variables, etc., is inherently more complex than univariate data, and consequently the ways in which the data can, say, depart from a particular model are infinitely more varied.) Often the investigator may not know in advance which structural features of his data will be most susceptible to meaningful interpretation. For example, he may apply some form of factor analysis to his data in a search for interpretable 'factors' in cases where some other aspect of the structure – such as the way individuals cluster into several homogeneous groups – would be a far more useful discovery. In such cases it is clear that to leave the analysis to any rigidly prespecified method, which must rest upon some specific assumption (whether explicit or implicit) about the data, is not what is required. In many situations the techniques which *are* needed for an investigation of multivariate data are informal

2

exploratory ones rather than formal *confirmatory* ones such as tests of specific hypotheses. Graphical techniques are well suited for this purpose since they tend to be less formal and confining, and enable the investigator to gain insight into the structure of his data without imposing restrictive constraints.

Computer Aids to Graphical Techniques

The usual methods of multivariate analysis require, in general, a large amount of computation, and their routine application in practice would be quite impossible without the aid of the electronic computer. For the methods to be described in this text the computer will usually be needed to perform the numerical computations involved, but in addition it may also be an extremely useful aid in producing automatically the graphical output arising from some particular technique. In this text we shall concentrate on methods for which this will in most cases involve simply a line printer plot or diagram, but occasionally, when more accurate outputs are required, a device such as a microfilm plotter may be necessary. Of course, far more sophisticated devices such as interactive graphics terminals, light pens, and other exotic hardware are available which may prove very exciting and useful tools for the investigation of data. For example, the interactive, conversational aspect of a computer allows the investigator to convey rapidly his needs to the machine, a process which would obviously be extremely advantageous in the analysis of complex multivariate data. This text will, however, be primarily concerned with techniques *not* requiring such devices as a necessity. The emphasis here will be on relatively simple techniques which do not require specialized hardware for the graphical display.

Plan of Text

This text is primarily directed towards research workers in such areas as psychology, psychiatry, medicine, and the social sciences. For this reason the 'technical details' sections, describing the mathematics of each method discussed, have been kept at a fairly low level requiring only a familiarity with elementary statistics and some knowledge of matrix algebra.

The numerical examples given for each of the methods may appear, at first sight, to involve 'small' data sets and many readers may complain that the size of these is unrelated to the size of problems he or she is normally concerned with in practice. However, the choice of *which* numbers to examine by one of the methods is open to consideration, and in many of the examples given the numbers used resulted from initial analyses of a *large* data set. This is not to say that large data sets present *no* difficulties, and the problem is considered further in the final chapter, page 96.

Since many of the techniques to be described operate not directly on the raw data matrix, **X**, but on a matrix of *similarities, dissimilarities,* or *distances*, an appendix is included (Appendix A) giving a brief description of such measures.

A further appendix (Appendix B) gives details of sources of various computer programs which either implement or are useful in implementing many of the methods discussed.

Summary

The techniques to be described, which are essentially graphical in nature, are useful for the informal examination of many types of multivariate data. They may be regarded as supplementing the more usual methods of multivariate analysis, and may prove useful in indicating which numerical calculations are relevant for further study, and for assisting with the interpretation and communication of results. Their use, in association with the more formal methods of multivariate analysis, should give added insight into the complex structure that may be present in much multivariate data. It should be remembered, however, that they in no way replace the more formal numerical techniques that may be available. Indeed there is need in statistical data analysis for both numerical *and* graphical techniques. The numerical techniques serve as objective yardsticks against which one can evaluate the degree of evidence contained in graphical displays. Tukey (1970) describes the relationship between the two approaches by the following analogy with the relationship between a police detective and a judge.

'The graphical techniques are the counterpart of the detective looking for any clues to help uncover the mysteries of the data. The formal numerical inference tools of classical statistics are the counterpart of the judge who weighs the degree of evidence contained in the clues to determine how much confidence to put into them.'

4

2. Ordination Techniques

Introduction

A visual examination of multivariate data could be made extremely easily if it involved only two variables, since each bi-variate observation could be represented simply as a point in the two-dimensional scattergram for the two variables. With three variables a three-dimensional model representing the data could be constructed. (Fraser and Kovats (1966) show how stereoscopic models of three-dimensional figures may be produced.) In practice, of course, the number of variables involved will be considerably greater than three, and consequently such direct graphical methods are not available.

It might be imagined, however, that scattergrams for *all* pairs of variables might be examined as a simple method of 'looking' at the data, but although this approach may be useful in some situations it is, in general, very unsatisfactory for two reasons. Firstly, if the number of variables is greater than about ten then the number of such plots to be examined is large, and such examination is as likely to lead to confusion as enlightenment about the structure of the data. Secondly, such plots may be very misleading since any structure present in the original p-dimensional space of the data is not necessarily reflected by that present in scattergrams of pairs of variables. This is a consequence of the well-known result that marginal distributions of multivariate data do not necessarily reflect the joint multivariate distribution of the variables. Examples of data illustrating this fact are given by Cattell and Coulter (1966), and by Nathenson (1971).

The question arises therefore that since we cannot, satisfactorily, use the original variables *directly* to obtain two-dimensional representations of our data, is it possible to use them *indirectly* to obtain such a representation? In this chapter several techniques which attempt this will be described. Such techniques, which are generally known as *ordination* or *mapping* methods, seek to obtain a representation of the original p-dimensional data in a reduced number of dimensions, say p^*, whilst maintaining as far as possible in some sense the structure present in the original space. (We shall, of course, be primarily concerned in representations with $p^* = 2$.)

Some of these techniques operate directly on the $(n \times p)$ data matrix $\mathbf{X}$; others on an $(n \times n)$ matrix of inter-individual distances or similarities which *may*, in some cases, have been obtained from **X** using one of the distance or similarity measures described in Appendix A, but may also in some situations have arisen directly, as we shall see in the examples given on pages 19 and 29. Before coming to ordination methods proper, however, we shall describe some simple methods whereby more than two variable values may be accommodated on a two-dimensional scattergram.

Representing Further Dimensions on a Two-dimensional Plot

There are several ways in which further variable values may be accommodated simply on a two-dimensional scattergram. For example, Gower (1967) reports a method due to Ross in which the first two variable values for each observation are plotted in the usual way to give, for a particular observation, a point R, say. The third variable value for this observation is now represented as a line with length proportional to this value originating from R and directed eastwards if the value is positive and westwards if negative. Similarly the fourth dimension uses the north and south directions, and if five and six dimensions are required the N.E./S.W. and N.W./S.E. directions could be used. Figure 2.1 shows a set of four-dimensional data represented by this technique.

Ball and Hall (1970) use a similar technique in their interactive graphic computer system. Again length of the line is used to represent the third variable, but these authors use the tilt of the line to represent the fourth variable.

When using an automatic plotting device such as a microfilm plotter a method which suggests itself for representing a third dimension on the two-dimensional scattergram is to alter the *intensity* of the plotted character. Faintly plotted characters can be used to represent observations lying far away on the z-axis, and brightly plotted characters those lying closer to the observer along this axis. Figure 2.2 shows such a plot. It is clear that the observations fall essentially into two groups differentiated mainly along the z-axis.

Representations such as these are restricted to a maximum of perhaps four or five dimensions and would obviously become confusing if there were too many points to plot. Consequently they are of limited importance in practice for dealing with the raw data. However, they might prove useful for displaying the *results* of some of the techniques to be discussed in the remainder of this chapter, as will be indicated in subsequent sections.

6

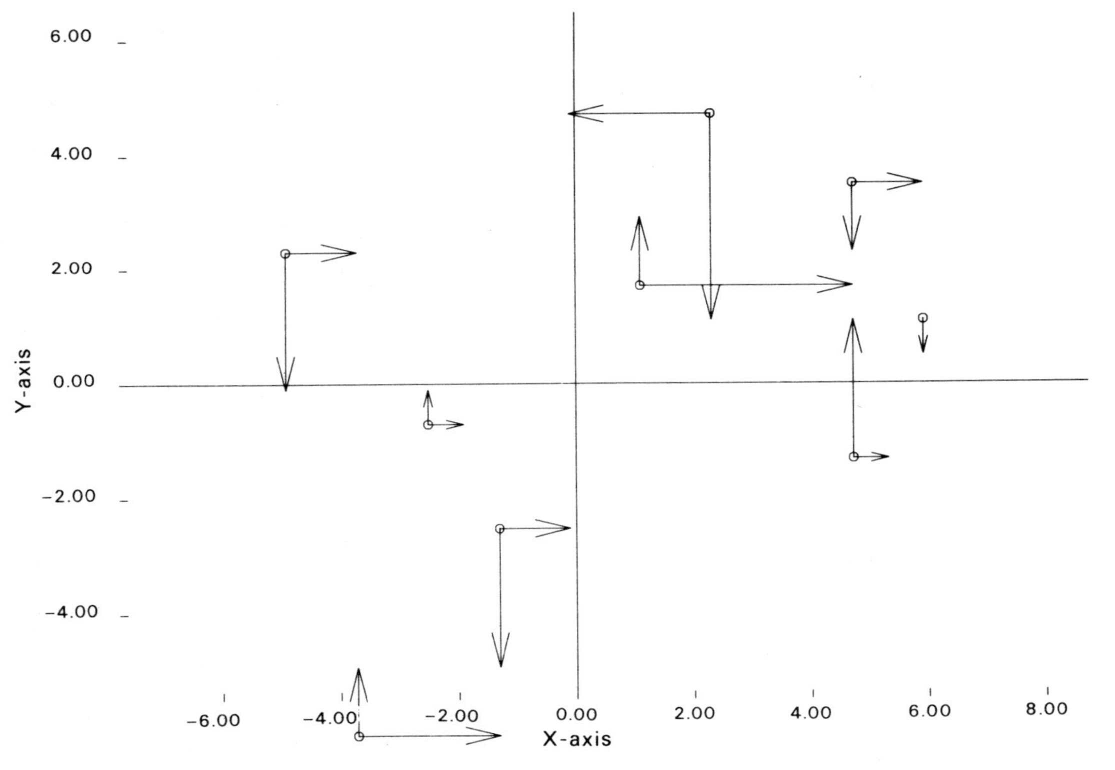

Figure 2.1   Four-dimensional data plotted on a two-dimensional diagram.

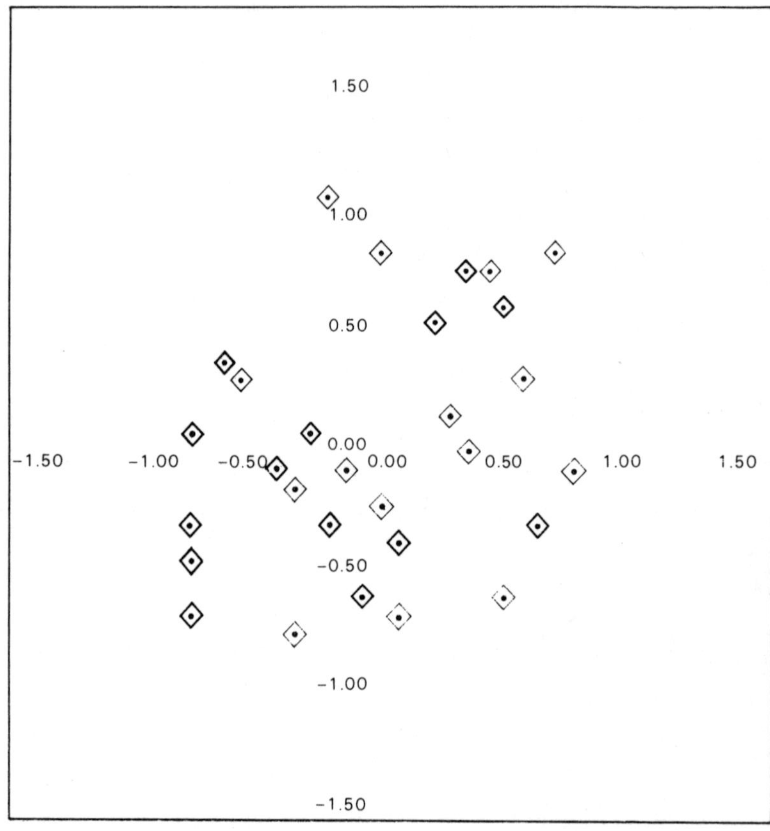

Figure 2.2   Three-dimensional data plotted on a two-dimensional diagram.

The description of the ordination techniques which are the main concern of this chapter begins with an account of how the well-known technique of principal components analysis may be used for a visual investigation of multivariate data.

Principal Components Analysis

Principal components analysis is described in detail in many textbooks of multivariate analysis (e.g. Press (1972)) and briefly below. We shall discuss its use as a method for obtaining a low-dimensional (specifically two-dimensional) representation of multivariate data so that the data

8

may be examined visually and any structure identified. The possibility of associating definite properties, e.g. 'factors', with the reduced number of dimensions will not be considered.

Principal components – technical details

Essentially, principal components analysis consists of finding linear transformations, $y_1, y_2, \ldots, y_p$, of the original variables, $x_1, x_2, \ldots, x_p$, that have the property of being uncorrelated. The y-variables are chosen in such a way that y_1 has maximum variance, y_2 has maximum variance subject to being uncorrelated with y_1, and so on. The transformation is obtained by finding the latent roots and vectors of either the covariance or the correlation matrix. The latent roots, arranged in descending order of magnitude, are equal to the variances of the corresponding y-variates, these being, of course, the principal components. Often the first few components account for a large proportion of the total variance of the x-variates and may therefore be used to summarize the original data.

An alternative method of viewing principal component analysis, and one which is perhaps preferable for the purpose of this text, is that given by Gower (1967) who shows that the first principal component may be regarded as the line of best fit (in the least squares sense) to the n,p-dimensional observations in the sample. These observations (which we shall henceforth denote by P_i, $i = 1, 2, \ldots, n$) may therefore be represented in one dimension by taking their projection, $\mathbf{Xl}_1$, onto this line, where $\mathbf{l}_1$ is the $(p \times 1)$ latent vector of the covariance (correlation) matrix, corresponding to the largest root. This representation will be perfect if the P_i are collinear but will be less good the more dimensions they account for. An improved representation is given by the projection of the observations onto the plane defined by the first two latent vectors of the covariance or correlation matrix. Similarly the first p^* latent vectors give the best fit in p^*-dimensions. In each case the original observations, P_i, may be projected into the lower dimensional space to give a set of points, say Q_i, the configuration of which acts as an approximation to the configuration of the observations P_i in the original p-dimensional space. How well the p^*-dimensional projection approximates the original p-dimensional configuration of the n observations is measured by the proportion of the variance in the data accounted for by the first p^* principal components. Since the total variance is given by trace (**S**), where the trace of a matrix is simply the sum of the elements in its main diagonal and **S** is the variance–covariance matrix, and the variance of each component is given by its corresponding latent root, this proportion is given by

$$P = \sum_{i=1}^{p*} \lambda_i / \text{trace (S)} \qquad\qquad \textbf{2.1}$$

where λ_i is the i-th latent root of S in descending order of magnitude.

We see, therefore, that if the first *two* principal components account for a large proportion of the total variance, then projecting the sample onto this plane should give a two-dimensional representation of the data which reflects fairly reasonably the structure present in the original p-dimensions. In cases where the first two principal components are not thought to account for a large enough proportion of the variance, then projections onto the first three or four components may be accommodated in a two-dimensional diagram by the type of procedure described in the preceding section. Judgement as to what value should be regarded as a 'large enough' proportion of the variance *may* be made with the assistance of a formal significance test on the latent roots of the covariance matrix; *see* Lawley and Maxwell (1971), Chapter Three. In general, however, such judgements will be made informally, with values of the order of 75 to 80 per cent or above being regarded as indicative of an adequate fit.

To summarize we may say that a principal components analysis transforms the original set of points P_i into a set of points Q_i. These two sets of points represent exactly the same geometrical configuration but are related to different sets of orthogonal axes. When only the first few principal component axes are used then the relative positions of the projected points Q_i are an approximation to the relative positions of the points P_i. The distance between two projected points Q_i and Q_j is an approximation to the *Euclidean distance* (see Appendix A) between the original sample observations P_i and P_j. (If projections are taken onto the complete set of principal component axes then the distances would, of course, be equal.)

Let us now examine some examples of using principal components to obtain a visual representation of multivariate data.

Principal components analysis applied to Fisher's iris leaf data
No text concerned with the analysis of multivariate data would, it seem, be complete without some reference to Fisher's iris data (Fisher (1936)). The present text is no exception, and the data, which consist of four measurements on each of fifty plants from three species of iris, are here subjected to a principal components analysis. We shall assume for the purposes of this example that we are exploring these data

without any *a priori* knowledge of their structure; i.e. we shall assume that we are unaware *a priori* of the three different species of iris present. The first two components account for 96 per cent of the total variance, and consequently projecting the data onto the plane defined by these components should give a very reasonable approximation to the original four-dimensional structure. Figure 2.3 shows the resulting two-dimensional plot.

This diagram clearly indicates the presence of 'clusters' in these data, although only one species of iris, namely *iris setosa*, is distinctly differentiated. It would be difficult from this diagram to infer the presence of the remaining two species, *iris virginica* and *iris versicolour*. However, the two-dimensional configuration obtained would indicate that some type of cluster analysis might usefully be applied to these data.

Using principal component plots to identify multivariate outliers
The consequences of having 'wild' observations in a multivariate sample are intrinsically more complex than in the much discussed univariate case (see, for example, Anscombe (1960)). One reason for this is that a multivariate outlier can distort not only measures of location and scale but also those of orientation, i.e. correlation. A further reason is the variety of types of multivariate outlier which may arise; e.g. a multivariate observation may be an 'outlier' because of a gross error in one of the variables or because of systemic mild errors in all of them. Gnanadesiban and Kettenring (1972) conclude that because of the complexity of the multivariate case it would be fruitless to search for a truly omnibus multivariate outlier detection procedure, and suggest that a more reasonable approach is to tailor detection procedures to specific types of situation. Such an approach recognizes that an outlier for one purpose may not necessarily be one for another purpose! Here we shall illustrate how principal component plots may be useful for detecting two types of multivariate outlier.

The first type are those which are inappropriately inflating variances and covariances (if working with the covariance matrix) or correlations (if working with the correlation matrix). Multivariate outliers of this variety may, in many cases, be detected by plotting the data in the space of the first two principal components. Figure 2.4. shows such a plot for a set of five-dimensional data. The presence of an outlier is clearly indicated, and this observation would need to be examined to decide whether or not it should be removed before attempting further analyses.

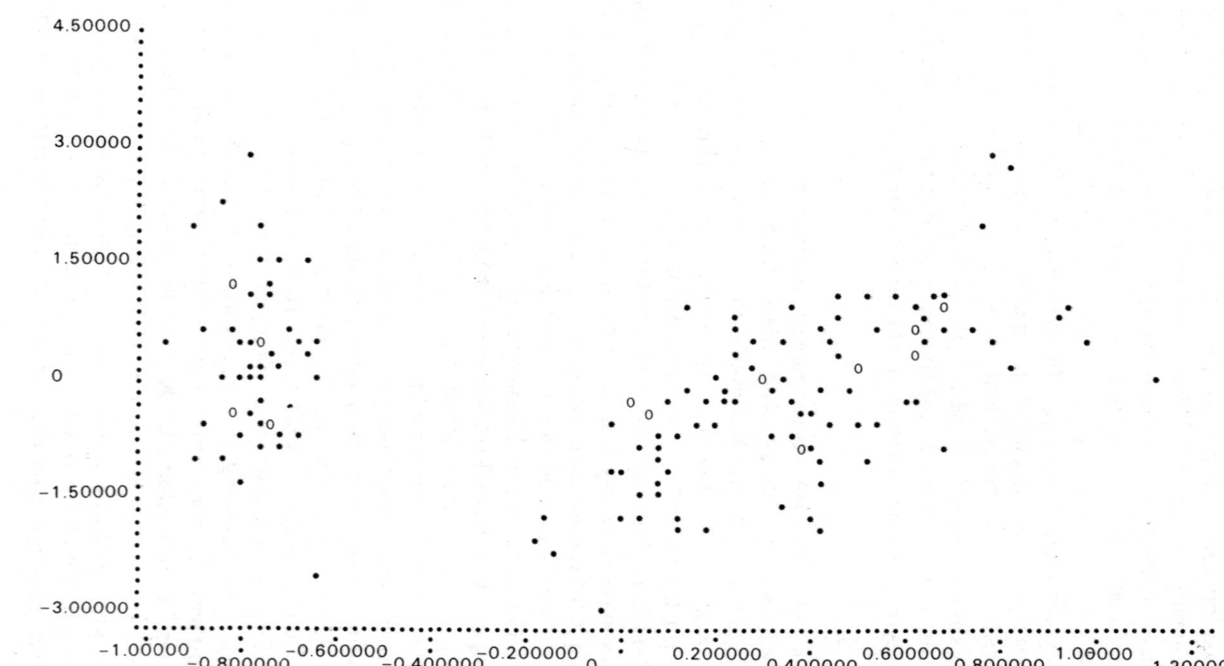

Figure 2.3   Fisher's iris leaf data plotted in the space of the first two principal components.

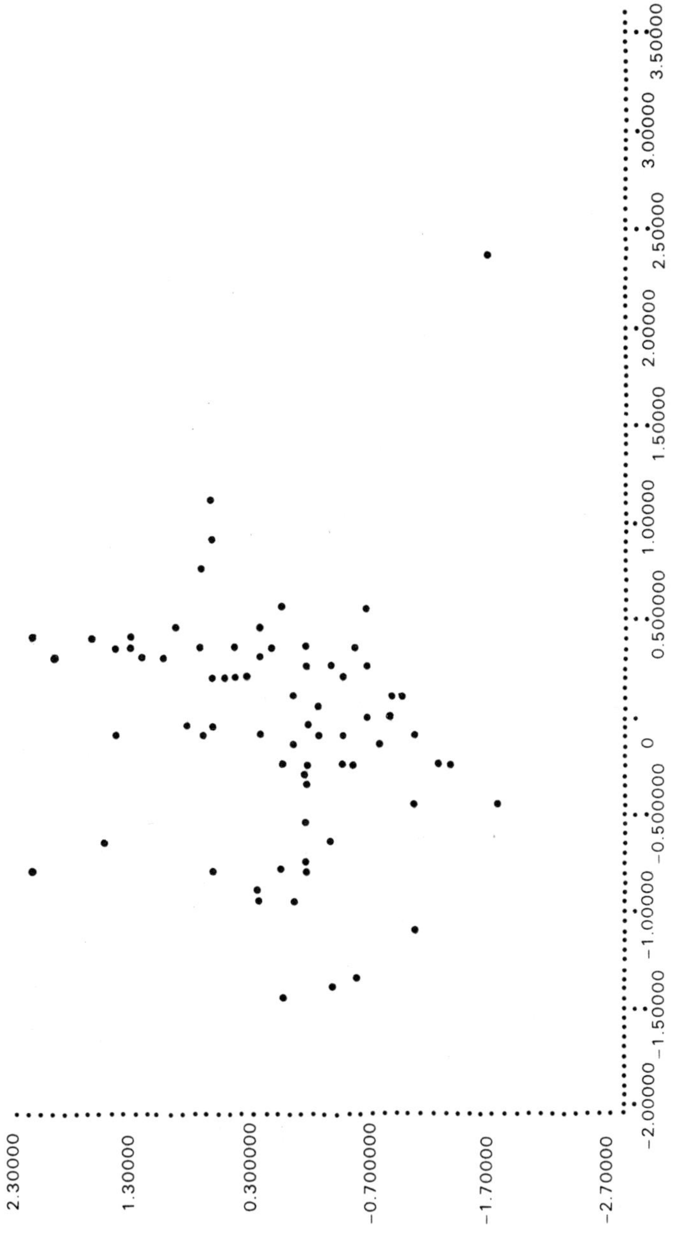

Figure 2.4 First example of using principal components plots to detect outliers. (Plot is in the space of the *first* two principal components.)

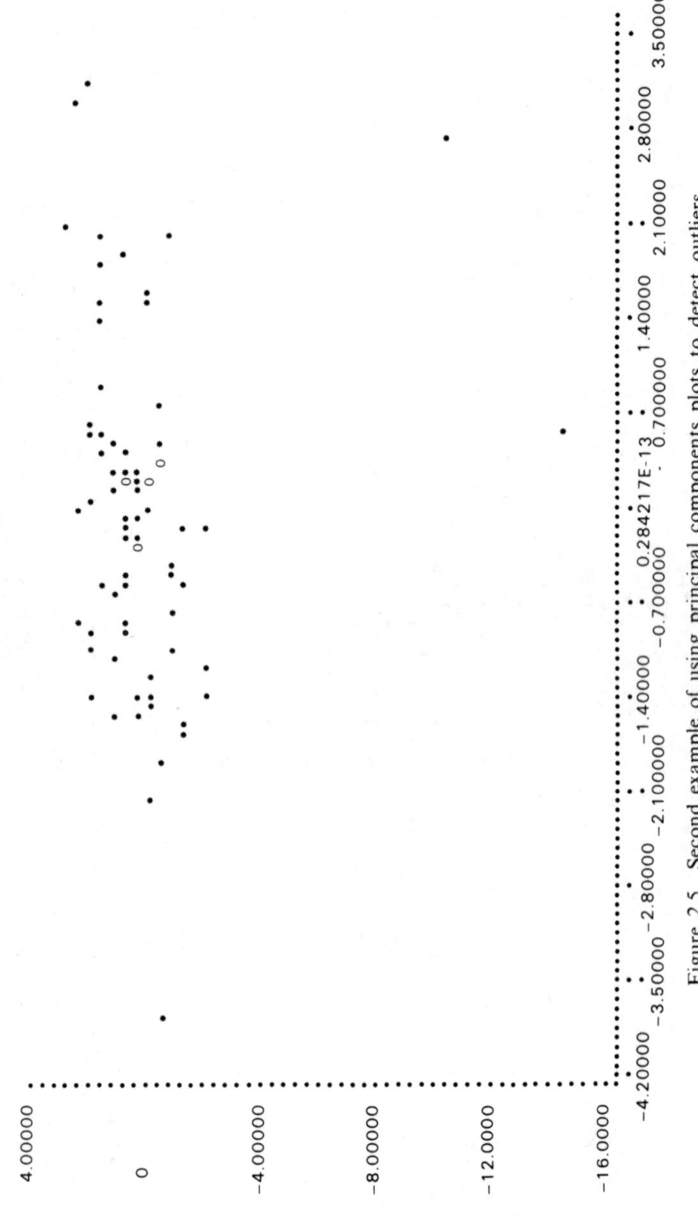

Figure 2.5   Second example of using principal components plots to detect outliers. (Plot is in the space of the *last* two principal components.)

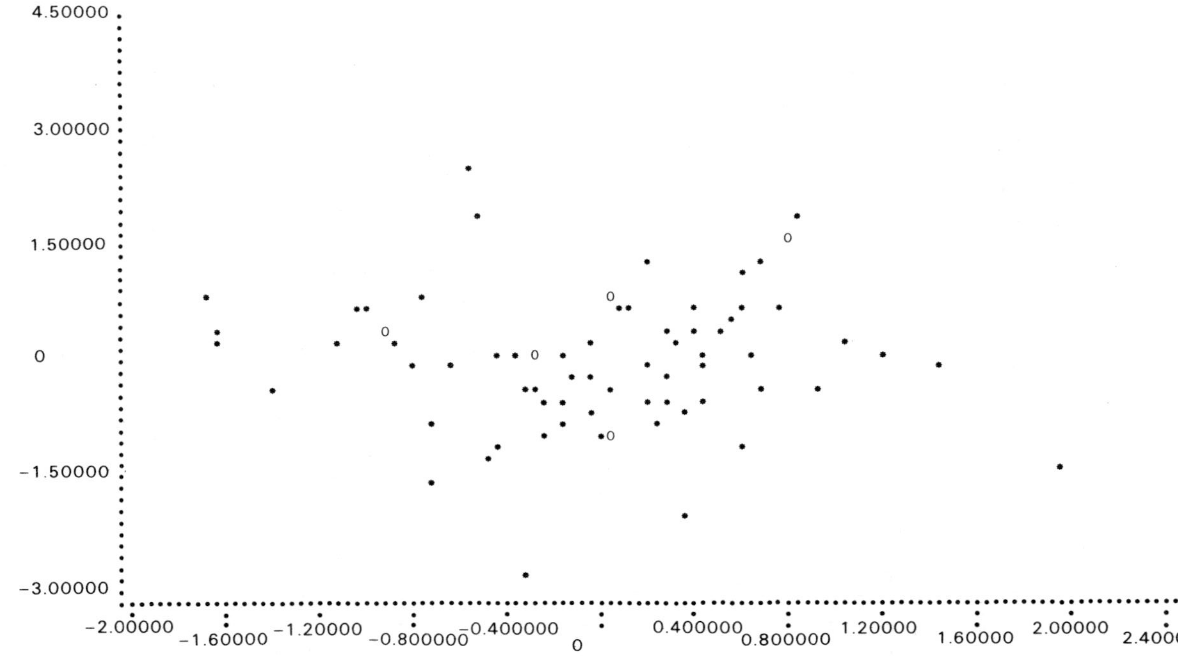

Figure 2.6   'Outlier' data from Figure 2.5 plotted in space of *first* two principal components.

A second type of outlier, which may usually be detected by plots in the space of the *last* few principal components, is the kind which is adding insignificant dimensions or obscuring singularities in the data. A set of five-dimensional data was constructed to contain two such outliers, and Figure 2.5 shows the two-dimensional plot of the data using principal components four and five. This diagram shows clearly the presence of the two outliers; repeating the analysis with these observations removed results in a solution in which only four principal components are needed to account for the variation in the data, indicating that for the remaining observations one of the variables is simply a linear combination of some or all of the others.

It is interesting to compare Figure 2.5 with a plot of these data in the space of the first two components, i.e. Figure 2.6; this shows no evidence of any outlying observations.

Some very interesting applications of using principal component plots to identify multivariate outliers are given in Gnanadesikan and Kettenring (1972).

Principal Co-ordinates Analysis

In a principal components analysis the distance between the projected points Q_i and Q_j is an approximation to the *Euclidean* distance between the original observations P_i and P_j. Gower (1966) describes a procedure which he terms principal co-ordinates analysis, which is similar in some respects to principal components analysis, but leads to a set of projected points Q_i whose distances apart may be allowed to reflect *non-Euclidean* distances between the original observations. In situations where the appropriate distance measure between observations is considered to be something other than Euclidean this would obviously be a very useful procedure.

The following account of this method follows that given by Gower.

Principal co-ordinate analysis – technical details
Let $\mathbf{A}$ be a symmetric $(n \times n)$ matrix with latent roots $\lambda_1, \lambda_2, \ldots, \lambda_n$ and associated $(n \times 1)$ latent vectors $\mathbf{c}_1, \mathbf{c}_2, \ldots, \mathbf{c}_n$ as shown in Table 2.1.

16

Table 2.1

Latent roots and vectors of the symmetric matrix $\mathbf{A}$

		λ_1	λ_2	λ_n
	Q_1	c_{11}	c_{12}	c_{1n}
Point	Q_2	c_{21}	c_{22}	c_{2n}
	$\vdots$	$\vdots$	$\vdots$	$\vdots$
	Q_n	c_{n1}		c_{nn}

(The elements of $\mathbf{c}_i$ are $c_{1i}, c_{2i}, \ldots, c_{ni}$.)

Suppose now we take the elements of the i-th row of Table 2.1 as the co-ordinates of a point Q_i in n-dimensional space. The Euclidean distance, Δ_{ij}, between points Q_i and Q_j in this space is therefore given by

$$\Delta_{ij}^2 = \sum_{r=1}^{n} (c_{ir} - c_{jr})^2$$

$$= \sum_{r=1}^{n} c_{ir}^2 + \sum_{r=1}^{n} c_{jr}^2 - 2 \sum_{r=1}^{n} c_{ir}c_{jr} \qquad \textbf{2.2}$$

If the latent vectors are normalized so that the sums of squares of their elements are equal to their corresponding latent roots, i.e. so that

$$\sum_{i=1}^{n} c_{ir}^2 = \lambda_r,$$

then it is well known that

$$\mathbf{A} = \mathbf{c}_1\mathbf{c}_1' + \mathbf{c}_2\mathbf{c}_2' + \ldots + \mathbf{c}_n\mathbf{c}_n' \qquad \textbf{2.3}$$

and therefore that

$$a_{ii} = \sum_{r=1}^{n} c_{ir}^2 \quad \text{and} \quad a_{ij} = \sum_{r=1}^{n} c_{ir}c_{jr}.$$

Using these results equation 2.2 becomes

$$\Delta_{ij}^2 = a_{ii} + a_{jj} - 2a_{ij} \qquad \textbf{2.4}$$

Suppose now that the Matrix $\mathbf{A}$ had elements $a_{ij} = -\frac{1}{2}d_{ij}^2$ and $a_{ii} = 0$, where d_{ij} is some measure of inter-individual distance. From equation

2.4 we see that Δ_{ij} is now simply equal to d_{ij}, and consequently the above procedure gives a method of finding co-ordinates for a set of points given their inter-distances d_{ij}. In particular if d_{ij} was Euclidean distance this method would be directly analogous to principal components analysis. However, the advantage of this method is that it may be used to find a set of co-ordinates for observations where the d_{ij}s are not considered to be Euclidean.

If $\mathbf{A}$ was a similarity matrix so that elements a_{ii} were unity then

$$\Delta_{ij}^2 = 2(1 - a_{ij})$$ 2.5

and principal co-ordinates analysis would lead to a spatial representation of the similarities in which Δ_{ij} of equation 2.5 functioned as Euclidean distance. (Not all similarity measures would be suitable, as we shall see later.)

We have now established that given a symmetric matrix $\mathbf{A}$ with elements a_{ij}, we may find a set of co-ordinates in n-dimensional space such that the Euclidean distance between the points in this space is given by Δ_{ij} in equation 2.4, and that this procedure may be used to find the co-ordinates of a set of observations given their inter-distances (*not* necessarily Euclidean), or their similarities. Gower shows further that it is legitimate to use principal components on these co-ordinates to find the best fit in fewer dimensions. The whole process therefore involves two stages, each stage requiring the determination of the latent roots and vectors of an $n \times n$ matrix. (That is, at stage one the matrix $\mathbf{A}$, and at stage two the $n \times n$ matrix of n-dimensional co-ordinates resulting from the first stage.) Gower shows that these two stages may be collapsed into one as follows:

1. Calculate the matrix $\mathbf{A}$. (In the case of similarities, $\mathbf{A}$ is simply the inter-individual similarity matrix; with distance measures, $\mathbf{A}$ would be formed by taking $a_{ii} = 0$ and $a_{ij} = -\frac{1}{2}d_{ij}^2$.)

2. Transform this to a matrix $\boldsymbol{\alpha}$, the elements of which are given by

$$\alpha_{ij} = a_{ij} - \bar{a}_{i.} - \bar{a}_{.j} + \bar{a}_{..}$$ 2.6

where $\quad \bar{a}_{i.} = \dfrac{1}{n}\sum_{j=1}^{n} a_{ij}, \ \bar{a}_{.j} = \dfrac{1}{n}\sum_{i=1}^{n} a_{ij}, \text{ and } \bar{a}_{..} = \dfrac{1}{n^2}\sum_{i=1}^{n}\sum_{j=1}^{n} a_{ij}.$

3. Find the latent roots and vectors of $\boldsymbol{\alpha}$, scaling each vector so that the sum of squares of its elements is equal to its corresponding latent root.

18

The elements of the k-th latent vector now give the co-ordinates of the n points on the k-th principal axis. The first two co-ordinates may now be used to obtain a visual representation of the original distance or similarity matrix. A measure of the adequacy of the fit of, say, the first $p*$ principal co-ordinates is given by

$$T = \sum_{i=1}^{p*} \gamma_i / \text{trace} (\alpha) \qquad\qquad \textbf{2.7}$$

where the γ_i are the latent roots of the matrix α arranged in descending order of magnitude.

Gower shows that this method may be used only with similarity measures which give rise to an α matrix with *no* negative latent roots; i.e. α must be positive semi-definite. The same author (Gower 1971a)) shows that this condition holds for a wide class of similarity measures.

Principal co-ordinates analysis is seen to have a considerable advantage over principal components analysis when seeking a visual representation of data. It operates directly on similarity and distance matrices and is not restricted to Euclidean distances. Principal components analysis will give a useful visual representation only when the distances between the original observations can safely be assumed to be Euclidean. The only advantage of this particular distance measure is that it allows the co-ordinates of the Q_i to be related *linearly* to the original variable values.

Principal co-ordinates analysis applied to a distance matrix
As an example of the use of principal co-ordinates analysis the method was applied to a matrix of distances between forty-eight British cities given in the Automobile Association handbook. These distances are road distances, and consequently are certainly not Euclidean or even metric. The resulting two-dimensional solution is shown in Figure 2.7. We see that this gives a fairly familiar outline!

Some distortion in the expected shape of the two-dimensional representation of these distances might be attributed to the use of road distances; e.g. the towns of the south-west are misplaced in relation to each other compared with their positions on the usual map of the British Isles. Dorchester, Plymouth, and Barnstaple are particularly poorly located. However, since the value of T (equation 2.7) for the first two principal co-ordinates for these data is 94.6 per cent, the two-dimensional fit can obviously be regarded as extremely good.

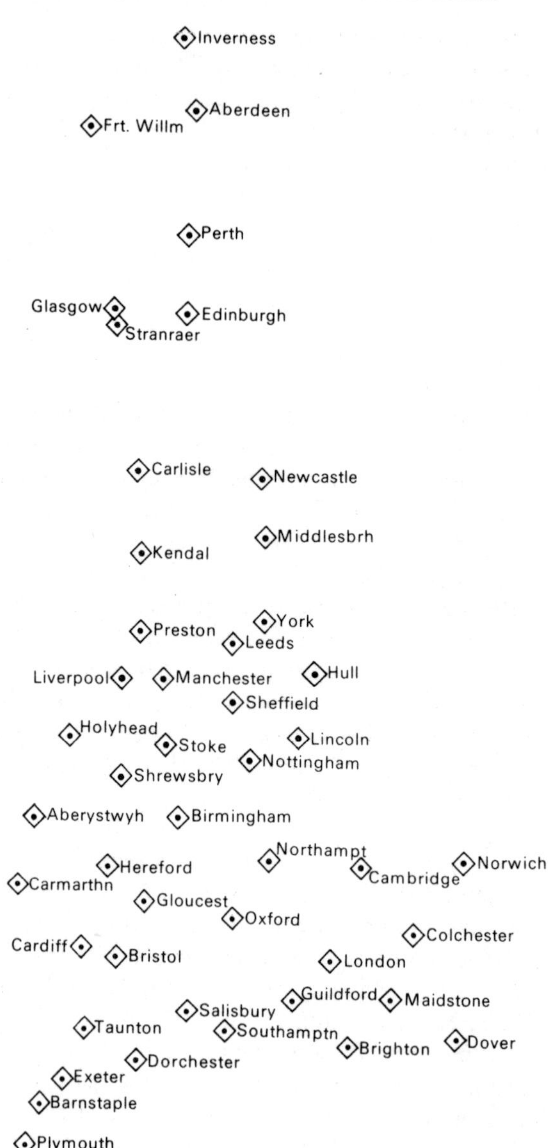

Figure 2.7   Two-dimensional solution given by applying principal co-ordinates analysis to matrix of distances between 48 British towns.

Principal co-ordinates analysis applied to a similarity matrix
In an investigation into various criteria which have been suggested for determining the diagnosis 'schizophrenia' amongst the mentally ill, a psychiatrist diagnosed 100 psychiatric patients (as schizophrenic or not), obtaining eight such diagnoses for each patient, each diagnosis being based on a different criterion. In addition, the hospital diagnosis and a diagnosis given by each of two computerized diagnostic procedures were available for each patient. The agreement between each pair of diagnoses for the 100 patients was then determined using Cohen's kappa coefficient (see Cohen (1960)). The resulting matrix of agreement coefficients is shown in Table 2.2.

Principal co-ordinates analysis applied to this matrix gives the two-dimensional solution shown in Figure 2.8. The value of the goodness of fit criterion, T, for this solution was 40.65 per cent, indicating only a reasonable fit. However, Figure 2.8 proved extremely useful to the psychiatrist involved in the study in enabling him to grasp the relationships between the various diagnostic criteria used. For example, we can see immediately that the two computer diagnoses are very different

Table 2.2

Similarity between various criteria for diagnosing schizophrenia

	1	*2*	*3*	*4*	*5*	*6*	*7*	*8*	*9*	*10*	*11*
1	1·00										
2	0·20	1·00									
3	0·26	−0·04	1·00								
4	0·11	0·30	0·08	1·00							
5	0·20	0·08	0·15	0·62	1·00						
6	0·27	0·22	0·20	0·50	0·53	1·00					
7	0·13	0·04	0·53	0·21	0·49	0·33	1·00				
8	0·07	0·28	0·15	0·54	0·41	0·58	0·18	1·00			
9	0·11	0·22	0·10	0·64	0·42	0·52	0·20	0·81	1·00		
10	−0·15	−0·05	−0·04	0·06	0·14	0·10	0·08	0·09	−0·11	1·00	
11	0·14	0·36	0·11	0·55	0·41	0·54	0·23	0·50	0·44	0·04	1·00

1 Taylor (T)	*2* Forrest Middle Life (FML)	*3* Forrest Young Adult (FYA)
4 Astrachan (A)	*5* Carpenter (C)	*6* Langfeldt (L)
7 Feighner (F)	*8* Schneider (S)	*9* Catego S (CS)
	10 Catego O P (COP)	*11* Hospital Diagnosis (HD)

(For details of the various diagnostic criteria used, etc., *see* Brockington and Leff (1976); the letters in brackets indicate the identification of diagnoses used in Figure 2.8.)

from each other, and that the 'Forrest middle life' criterion is well separated from all others. The plot also proved useful in communicating the results of this study to other psychiatrists.

Further examples of the use of principal co-ordinates analysis are available in Hills (1969) and Blakith and Reyment (1971).

The Biplot

Gabriel (1971) describes a procedure which he calls *the biplot* that allows a graphical display of the relationships between individuals, as indicated by certain measures of inter-individual distance, *and* between variables as indicated by their covariances and correlations. The biplot diagram also displays the variances of variables and allows the investigator to view the individual observations and their differences. The following technical description of the method is a summary of that given by Gabriel.

The biplot – technical details

This technique is based on the well-known result that any $n \times m$ matrix, $\mathbf{Y}$, of rank r can be factorized as follows:

$$\mathbf{Y} = \mathbf{GH}' \qquad\qquad 2.8$$

where $\mathbf{G}$ is an $n \times r$ matrix and $\mathbf{H}$ is an $m \times r$ matrix. Both $\mathbf{G}$ and $\mathbf{H}$ are necessarily of rank r. (This factorization is not, of course, unique.) The factorization given in 2.8 may be rewritten as

$$y_{ij} = \mathbf{g}_i'\mathbf{h}_j \qquad\qquad 2.9$$

for each i and j, where y_{ij} is the element of the i-th row and j-th column of $\mathbf{Y}$, $\mathbf{g}_i'$ is the i-th row of $\mathbf{G}$, and $\mathbf{h}_j'$ the j-th row of $\mathbf{H}$. In this form the factorization assigns the vectors $\mathbf{g}_1 \ldots \mathbf{g}_n$ one to each of the n rows of $\mathbf{Y}$, and the vectors $\mathbf{h}_1 \ldots \mathbf{h}_m$ one to each column of $\mathbf{Y}$. Each of these vectors is of order r, and consequently equation 2.9 indicates a method for representing the matrix $\mathbf{Y}$ by means of $n + m$ vectors in r-space. If $\mathbf{Y}$ is of rank two then these $n + m$ vectors may be plotted in the plane giving a representation of the nm elements of $\mathbf{Y}$ by means of the inner products of the corresponding row and column vectors. Such a plot is termed by Gabriel *a biplot*.

Matrices of rank higher than two cannot be represented exactly by a biplot. However, if the matrix $\mathbf{Y}$ can be satisfactorily approximated by a rank two matrix, $\mathbf{Y}_{(2)}$, then the biplot of $\mathbf{Y}_{(2)}$ may allow a useful

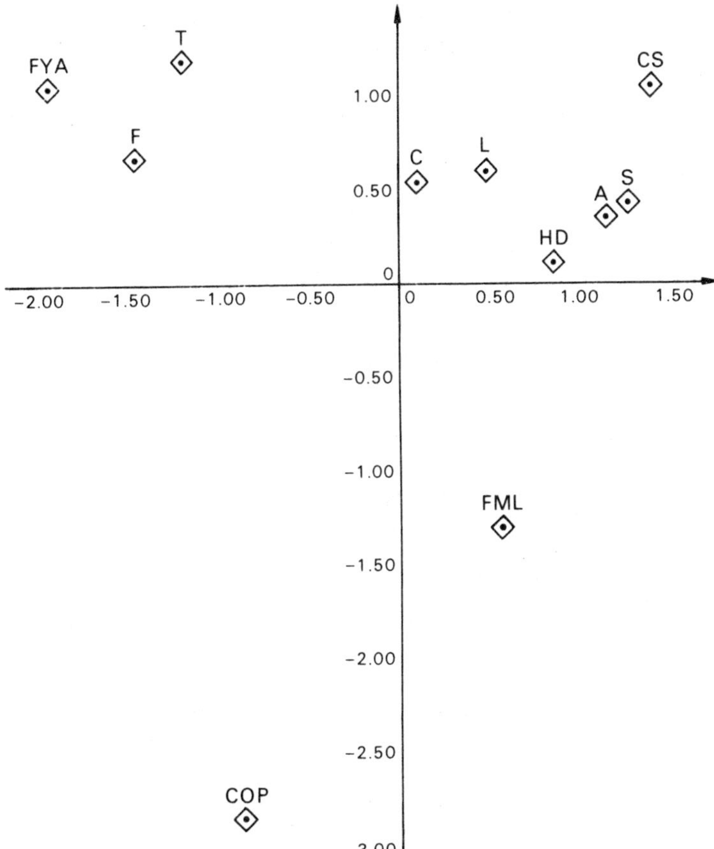

Figure 2.8   Two-dimensional solution given by applying principal co-ordinates analysis to the diagnostic similarity matrix shown in Table 2.2.

approximate visual inspection of Y itself. In the context of data analysis Gabriel shows that an extremely useful biplot is that obtained from consideration of the latent roots and vectors of the sample variance–covariance matrix, S; a rank two approximation to the original data matrix is obtained from these as follows:

$$Y_{(2)} = [p_1 p_2] \begin{bmatrix} \sqrt{\lambda_1} & 0 \\ 0 & \sqrt{\lambda_2} \end{bmatrix} \begin{bmatrix} q'_1 \\ q'_2 \end{bmatrix} \qquad \textbf{2.10}$$

where λ_1 and λ_2 are the first two latent roots of the matrix $n\,\mathbf{S}$ (where n is sample size), and $\mathbf{q}_1(p \times 1)$ and $\mathbf{q}_2(p \times 1)$ are the corresponding latent vectors.

The vectors $\mathbf{p}_1(n \times 1)$ and $\mathbf{p}_2(n \times 1)$ are given simply by

$$\mathbf{p}_i = \frac{1}{\sqrt{\lambda_i}} \mathbf{Y}\mathbf{q}_i \qquad\qquad \textbf{2.11}$$

where $\mathbf{Y}$ is the data matrix with the mean for each variable subtracted out. Gabriel now shows that the exact biplot of $\mathbf{Y}_{(2)}$ (i.e. the approximate biplot of $\mathbf{Y}$) with the following choice of $\mathbf{G}(n \times 2)$ and $\mathbf{H}(p \times 2)$,

$$\mathbf{G} = [\mathbf{p}_1, \mathbf{p}_2]\sqrt{n} \qquad\qquad \textbf{2.12}$$

$$\mathbf{H} = \frac{1}{\sqrt{n}}[\sqrt{\lambda_1}\mathbf{q}_1, \sqrt{\lambda_2}\mathbf{q}_2] \qquad\qquad \textbf{2.13}$$

has the following desirable properties:

1. The Mahalanobis distance (see Appendix A, page 101) between *individuals i and j* is approximated by

$$\|\mathbf{g}_i - \mathbf{g}_j\|^2 \qquad\qquad \textbf{2.14}$$

2. The covariance of *variables i and j* is approximated by

$$\mathbf{h}_i'\mathbf{h}_j \qquad\qquad \textbf{2.15}$$

3. The variance of the j-th variable is approximated by

$$\|\mathbf{h}_j\|^2 \qquad\qquad \textbf{2.16}$$

4. The correlation of the i-th and j-th variable is approximated by the cosine of the angle between vectors $\mathbf{h}_i$ and $\mathbf{h}_j$

($\|\mathbf{w}\|$ represents the length of the vector $\mathbf{w}$).

This particular choice of approximate biplot of the data matrix, $\mathbf{Y}$, therefore allows the investigator to scan the distances between individuals and to inspect the variances, covariances, and correlations of variables simply by examining the lengths of vectors and the angles between them. Consequently this should prove a very useful graphical aid in interpreting multivariate matrices of observations, provided, of course, that these can be adequately approximated at rank two. Of interest in this context are the following goodness of fit measures:

1. The elements of the data matrix $\mathbf{Y}$ are biplotted with goodness of fit

$$P_1 = (\lambda_1 + \lambda_2) \bigg/ \sum_{i=1}^{r} \lambda_i \qquad \textbf{2.17}$$

2. The elements of $\mathbf{S}$, the variance–covariance matrix, are plotted with goodness of fit

$$P_2 = (\lambda_1^2 + \lambda_2^2) \bigg/ \sum_{i=1}^{r} \lambda_i^2 \qquad \textbf{2.18}$$

3. The inter-individual distances are plotted with goodness of fit

$$P_3 = 2/r \qquad \textbf{2.19}$$

(Readers are reminded that r is the rank of the data matrix $\mathbf{Y}$; in most cases this will, of course, be equal to p, the number of variables.)

We see, therefore, that whereas the observations themselves and the variances, covariances, and correlations may be well represented by the biplot, the inter-individual distances may not be adequately fitted.

An alternative biplot may be obtained by choosing

$$\mathbf{G} = (\sqrt{\lambda_1}\mathbf{p}_1, \sqrt{\lambda_2}\mathbf{p}_2) \qquad \textbf{2.20}$$

and

$$\mathbf{H} = (\mathbf{q}_1, \mathbf{q}_2) \qquad \textbf{2.21}$$

This factorization would be appropriate if we considered the distance between individuals to be Euclidean rather than Mahalanobis. A disadvantage of this choice of $\mathbf{G}$ and $\mathbf{H}$ is that approximations to variances and covariances such as those given in formulae 2.15 and 2.16 are no longer valid.

An example of the application of this technique should help to make the above description somewhat clearer.

The biplot for a set of psychiatric data
To illustrate the biplot technique we shall apply it to a set of psychiatric data consisting of scores on each of six psychiatric syndromes for a sample of twenty-five mentally ill patients. Using the biplot based upon the latent vectors of the covariance matrix (i.e. formulae 2.12 and 2.13), we obtain the diagram shown in Figure 2.9. Using formula 2.18 we find

that the elements of the covariance matrix for these data are plotted with a goodness of fit of 99 per cent. Consequently the biplot should give a very good representation of the properties of the six variables. Examining Figure 2.9 we see that the lengths of the vectors representing the variables show at a glance that five of the variables have approximately equal variances, whilst the sixth has a considerably lower value for this statistic. In addition the diagram shows clearly that variables one and two are highly correlated as are variables three, four, and five. There would appear to be almost zero correlation between each of variables one and two and variables three, four, and five. The vectors representing the patients are clearly seen to fall into three groups, indicating that these data consist of three types of patient. There is also some evidence that one of the patients (number twenty-five) is in some way distinct from the remainder. It is clear that for these data the biplot has provided a useful and informative visual display of the characteristics of the variables *and* the individuals.

Non-metric Multidimensional Scaling

Many psychological experiments involve judgements as to the similarity of stimuli, concepts, persons, traits, cultures, etc.; each experiment gives rise to a matrix of similarities between the stimuli, etc., on which judgements are being made, and there has long been interest in methods for representing the structure underlying such matrices. Early work in this area was that of Torgerson (1952) who developed a method similar to that described on page 16; the method is generally referred to in psychology as 'classical' or 'metric' multidimensional scaling. Now data in these experiments usually come from human subjects or observers who can readily and reliably give only *ordinal* judgements (e.g. they can specify that one stimulus is 'larger' than another without being able to attach any value to the exact numerical difference between them). Consequently the similarity judgements do not have strict numerical significance, and there is rarely any justification for using the particular numerical values rather than values obtained from some *monotonic transformation*. Because of this a number of methods have been developed under the general heading of 'non-metric multidimensional scaling' which seek a co-ordinate representation of similarity matrices using only the ordinal properties of the similarity measures. The first such technique was proposed by Shepard (1962a, b). This method was formalized by Kruskal (1964a, b). Further develop-

26

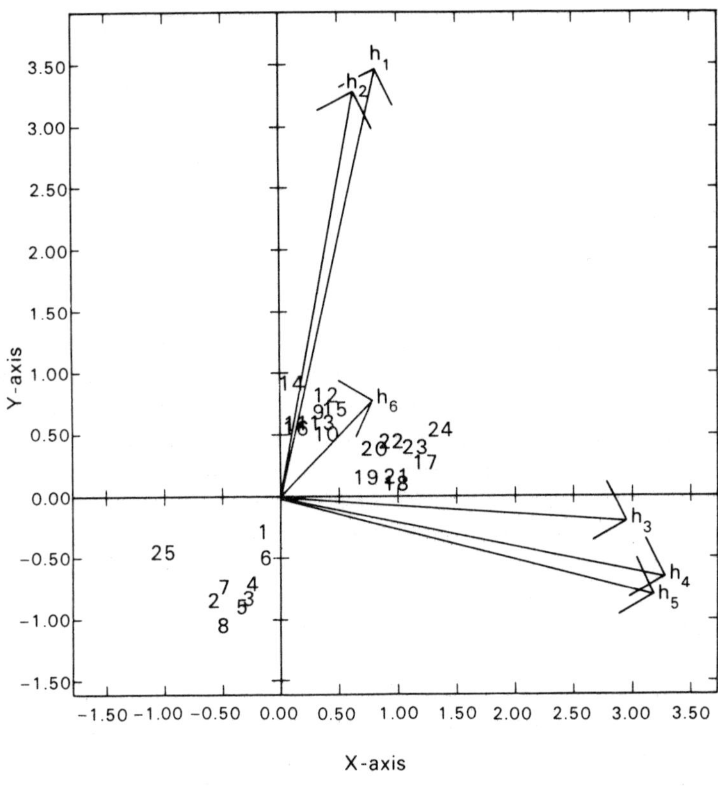

Figure 2.9 The biplot for a set of psychiatric data. (The end-points of vectors $\mathbf{g}_1$, $\mathbf{g}_2 \ldots \mathbf{g}_{25}$, are indicated by the numbers $1 \ldots 25$; the g-vectors have not here been plotted as lines.)

ments in this area have been numerous, and amongst the most notable are those by Coombs (1964), Carroll and Chang (1970), Guttman and Lingoes (1967), and Tucker (1972). Details of many of these techniques are to be found in Shepard *et al.* (1973) and Shepard (1974). Here we shall describe the most commonly used of these methods, namely the one originally proposed by Shepard and Kruskal.

Non-metric multidimensional scaling – technical details
Let us assume that the similarity matrix of concern is symetric with elements s_{ij}. Because of the symmetry we need only concern ourselves

27

with the $n(n-1)/2$ similarities in say the lower half of the matrix. Assuming that there are no ties, i.e. no two of the $n(n-1)/2$ similarities are the same, then it is possible to rank the similarities in strictly ascending order thus:

$$s_{i_1 j_1} < s_{i_2 j_2} < \ldots < s_{i_M j_M} \qquad \textbf{2.22}$$

where $M = n(n-1)/2$.

We wish to represent the n objects by n points in p^*-dimensional space, by finding a set of p^*-dimensional co-ordinates for each of them. As previously, let us denote the p^*-dimensional points by Q_i.

Suppose first that we had the points Q_i and that the corresponding Euclidean distance between points Q_i and Q_j was given by d_{ij}. We could then compare the similarities with these distances to see how well they matched in the sense of *large* distances corresponding to *small* similarities. A *perfect* match would mean that whenever one similarity is smaller than another, then the corresponding distances should satisfy the opposite relationship. In other words, a perfect match would mean that if we laid out the distances d_{ij} in a manner corresponding to formula 2.22 then the largest distance would come first and the other distances follow in descending order. That is,

$$d_{i_1 j_1} > d_{i_2 j_2} > \ldots > d_{i_M j_M} \qquad \textbf{2.23}$$

Therefore to obtain a perfect match, in this sense, between the similarities and the distances we should seek the points Q_i such that the Euclidean distances between them in the p^*-dimensional space are monotonically related to the similarities. In this way the numerical values of the s_{ij} are not of importance, only their rank order. Now for a given value of p^* the monotonicity property cannot, in general, be completely satisfied and some means is needed of assessing the extent to which a configuration falls short of the requirement. For this purpose Kruskal introduces a measure called *stress* and chooses the points Q_i so as to minimize this measure; stress is given by

$$S = \frac{\sum\limits_{i<j} (d_{ij} - \hat{d}_{ij})^2}{\sum\limits_{i<j} d_{ij}^2} \qquad \textbf{2.24}$$

In this formula the $\hat{d}_{ij}$ are a set of numbers known to be monotonic with the similarities. They are *not* distances; there is no configuration whose inter-point distances are $\hat{d}_{ij}$. They are merely a sequence of numbers, monotonic with the s_{ij}, used to reference the non-monotonicity of the distances d_{ij}. Examining formula 2.24 we see that 'stress' is essentially a 'residual sum of squares', familiar in many areas of statistics, merely normalized to make it invariant under changes of scale.

The problem of finding the p^*-dimensional configuration, Q_i, $i = 1 \ldots n$, which minimizes S, is essentially one of minimizing a function of many variables, the variables here being the co-ordinates of each point Q_i. There will, therefore, in this case, be a total of np^* variables to be determined by the minimization process. An efficient minimization procedure is described in Kruskal's second paper (Kruskal (1964b)). Basically, the procedure begins with an arbitrary configuration and then moves points around so as to minimize the stress value, continuing the procedure until a configuration is reached from which no improvement is possible. Roughly speaking, points Q_i and Q_j are moved closer together if $\hat{d}_{ij} < d_{ij}$ and apart in the opposite case to make d_{ij} more like $\hat{d}_{ij}$. As with all such minimization procedures, problems arise because of the presence of 'local minima'. One possible way around such a problem is to repeat the procedure employing a different initial configuration.

For this technique, S acts as a measure of the goodness of fit of the configuration of points obtained as a solution. Kruskal (1964a) offers the following informal evaluation of particular values of the stress.

Stress	*Goodness of fit*
20%	poor
10%	fair
5%	good
$2\frac{1}{2}$%	excellent
0%	perfect

(Perfect in this context simply means that there is a perfect monotone relationship between the similarities and the distances.)

Let us now examine an example of the application of this technique.

Non-metric multidimensional scaling applied to a matrix of 'distances' between famous Frenchmen

The set of data shown in Table 2.3 results from averaging the judgements of several individuals when asked to rate how alike they considered each pair in a set of ten famous (infamous!) Frenchmen. Ratings were given from unity, indicating that the pair were considered very similar, to nine, indicating complete dissimilarity. Table 2.3 may therefore be regarded as defining the 'distances' between the persons

29

concerned. Considering, however, the manner in which these distances were obtained it would seem sensible to regard them as having only ordinal significance; consequently non-metric multidimensional scaling was applied, resulting in the two-dimensional representation shown in Figure 2.10. The value of the stress for this configuration was 8.8 per cent, indicating a fairly reasonable fit. A knowledge of French politics is obviously an asset in interpreting this diagram; however, even from rather superficial knowledge of the men involved, the configuration given in Figure 2.10 does not seem unreasonable.

Many applications of this and related techniques have been made, and several of these are reported in the two volumes of papers edited by Shepard (see Shepard (1973)).

Non-linear Mapping

A further ordination technique first proposed by Sammon (1969) and known as *Non-linear mapping*, or NLM for short, will be described in this section beginning with the following brief technical description of the method.

Table 2.3

Ratings of distances between 'famous' Frenchman

	1	2	3	4	5	6	7	8	9	10
1	1									
2	4	1								
3	7	8	1							
4	8	8	2	1						
5	7	8	8	2	1					
6	1	6	8	8	7	1				
7	4	6	6	8	4	5	1			
8	5	8	6	7	3	3	5	1		
9	7	8	2	4	4	8	7	6	1	
10	6	8	3	3	4	8	5	6	2	1

1 Pompidou (P)	2 Vignancourt (V)	3 Duclos (D)
4 Geismar (G)	5 Mitterand (M)	6 De Gaulle (DG)
7 Poher (PR)	8 Schreiber (S)	9 Séguy (SY)
	10 Sartre (ST)	

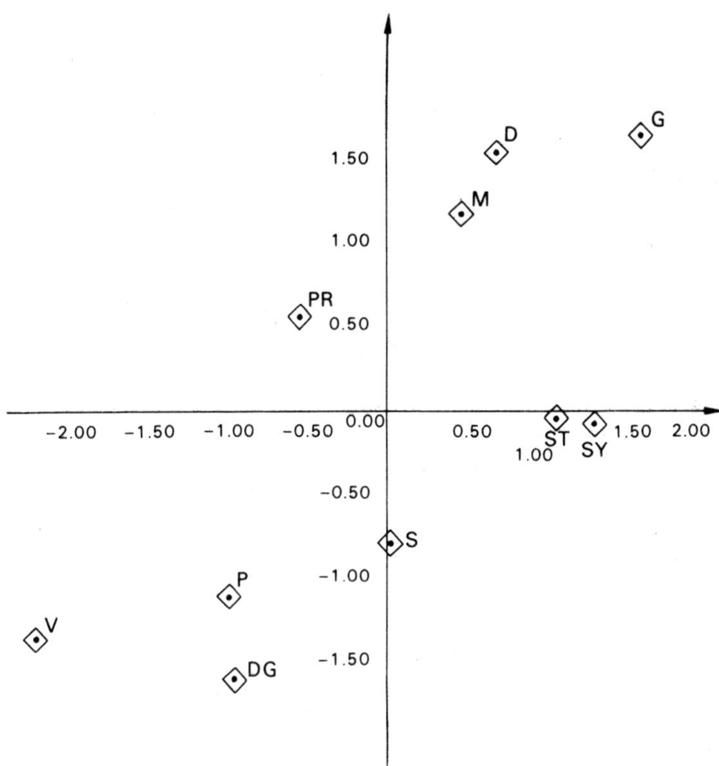

Figure 2.10   Two-dimensional solution given by non-metric multidimensional scaling applied to the matrix of 'distances' between famous Frenchman shown in Table 2.3.

Non-linear mapping – technical details
Non-linear mapping seeks to obtain a low dimensional representation of p-dimensional multivariate data by seeking a p^*-dimensional configuration which minimizes the expression E given by

$$E = \frac{1}{\sum\limits_{i<j} d_{ij}} \sum_{i<j} \frac{(d_{ij} - d_{ij}^*)^2}{d_{ij}} \qquad \textbf{2.25}$$

where d_{ij} is the Euclidean distance between observations i and j in the original p-dimensional space, and d_{ij}^* is the Euclidean distance between

31

the p^*-dimensional points representing these observations in the lower dimensional space. E, which is known as the *mapping error*, is, as with the Stress measure of the preceding section, a function of np^* variables, namely the co-ordinates of the points in the lower dimensional space; to minimize E, Sammon employs a 'steepest descent procedure' beginning with an arbitrary p^*-dimensional configuration. In the case of $p^* = 2$, the values of the two variables amongst the original p variables having largest variance are used to provide the starting configuration. Chang and Lee (1973) describe an alternative procedure for minimizing E, namely a 'relaxation' method (see for example, Allen (1954)). These authors also introduce a technique to cope with problems that arise for non-linear mapping, when there are a large number of sample points.

As with the stress measure of the preceding section, we see that E is simply a weighted sum of squares. The weighting used tends to preserve 'local' structure in the sense that each point in the p^*-space will bear approximately the same relationship to its near neighbour as it did in the original space, whilst its relationship to points relatively distant from it in p-space may be quite different in p^*-space.

Howarth (1973) suggests that the value of the mapping error associated with a configuration may be used to obtain information on the 'true' dimensionality of a given data set by employing some results of Elefante on the mapping error for a simplex. The mapping errors given by Elefante appear to be an upper bound for any data set whose dimensionality is the same as the simplex and thus indicate the true dimensionality of the data.

Let us now examine some examples of the application of this technique.

Non-linear mapping applied to set of data generated to contain clusters
Thirty, five-dimensional multivariate data points were generated by sampling ten observations from each of three five-variate normal distributions having different means, but the same variance–covariance matrix. The two-dimensional solution given by NLM applied to these data appears in Figure 2.11, and clearly indicates the presence of three clusters.

An application of non-linear mapping to a set of data generated to contain 'non-linear' structure
A further set of five-dimensional data was generated by adding random variation to the co-ordinates of points distributed evenly along a five-

32

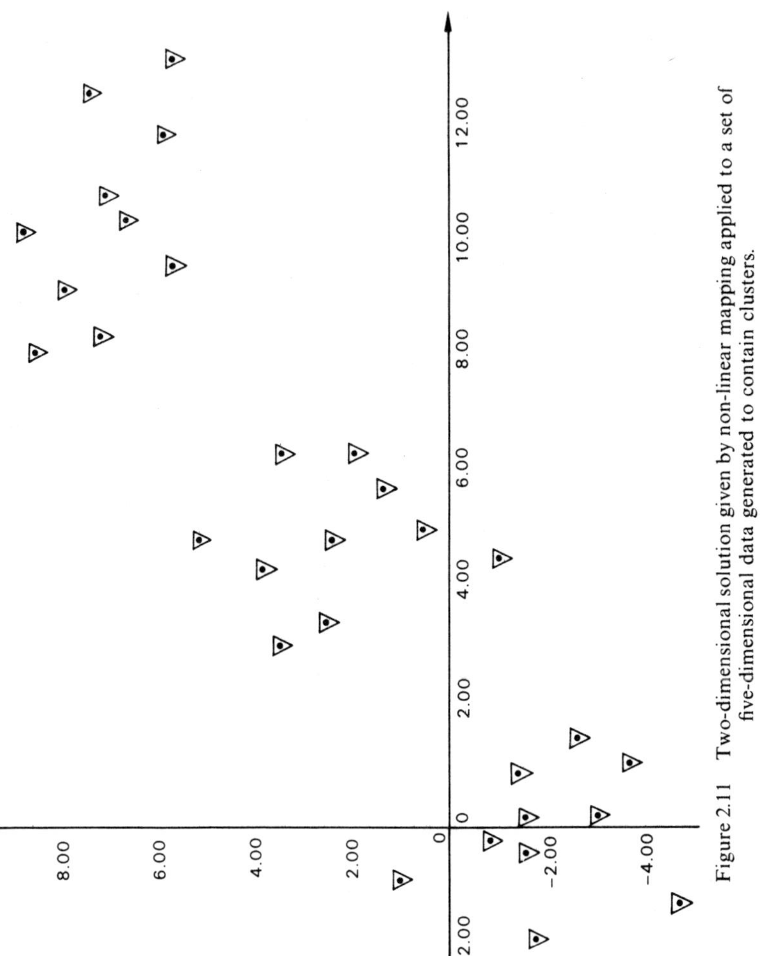

Figure 2.11   Two-dimensional solution given by non-linear mapping applied to a set of five-dimensional data generated to contain clusters.

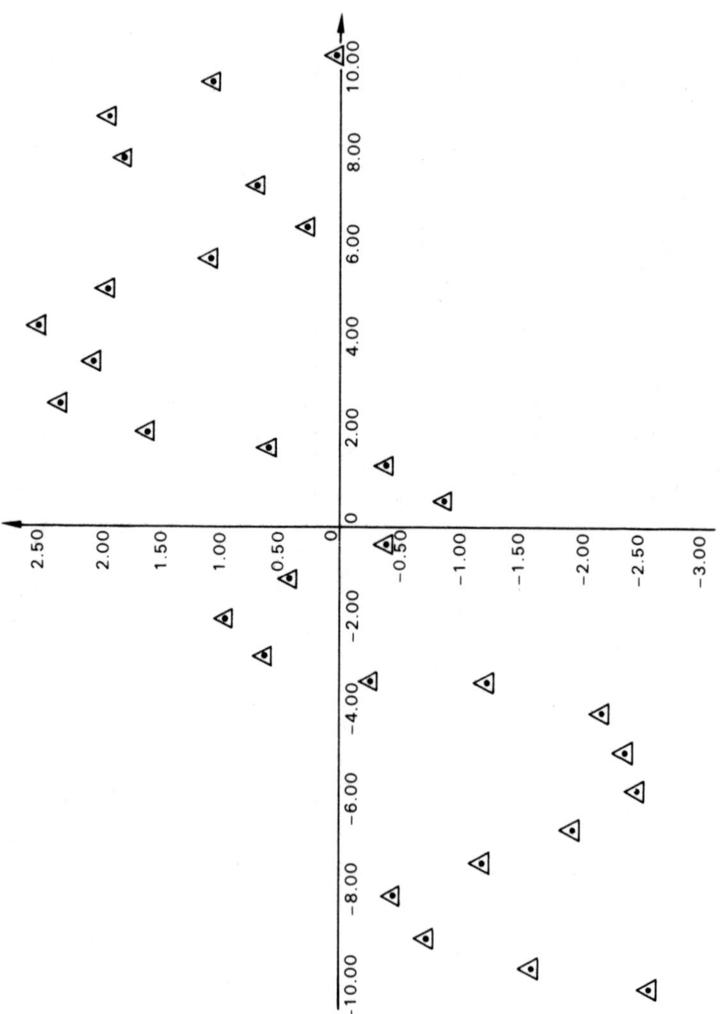

Figure 2.12 Two-dimensional solution given by non-linear mapping applied to a set of data lying along a curve in five dimensions.

dimensional curve. The parametric equations for this curve are as follows:

$$X = \cos(Z),\ Y = \sin(Z),\ U = 0.5\cos(2Z),\ V = 0.5\sin(2Z),\ = Z = \frac{1}{\sqrt{2}}t.$$

Thirty points were generated corresponding to values of t from 0 to 29. The two-dimensional solution resulting from the application of NLM to these data is shown in Figure 2.12. The inherent structure of these data is clearly indicated by this diagram.

Some interesting applications of non-linear mapping in the construction of a document classification space are described by Sammon. Chang and Lee apply their modification of the technique to a problem in automatic character recognition, and Howarth describes several applications of the method in geology.

Assessing the Goodness of Fit of Two-dimensional Representations of Multivariate Data

The usefulness of the techniques discussed in the previous section will be judged by how well the inherent structure in the data is preserved by the ordination procedure. Essentially this will depend on how well the original distances between the observations are maintained under the mapping process. In part this may be judged by the goodness of fit criteria already mentioned in connection with particular techniques. However, other methods are available for comparing two sets of distances which are often very useful. For example, Sokal and Rohlf (1962) and Kruskal and Carroll (1969) suggest correlating the $\binom{n}{2}$ distance pairs $(d_{ij},\ d_{ij}^*)$ where the d_{ij}s are the original distances between the observations and the d_{ij}^*s are the distances between points representing these observations and resulting from some particular ordination process. This correlation coefficient is generally known as the *cophenetic correlation*. A further coefficient for comparing two sets of distances proposed by Hartigan (1967) is δ given by

$$\delta = \sum_{i<j} |d_{ij} - d_{ij}^*| \qquad \textbf{2.26}$$

Gower (1970, 1971b) points out some problems with using the cophenetic correlation and describes an alternative geometrical approach to the problem. Gower's technique is used by Krzanowski (1971) to compare various distance measures applicable to multinomial data.

35

In this section we shall discuss a procedure which is particularly suitable for assessing how well the original distances are preserved by a *two-dimensional mapping*, and so in the context of this monograph is especially relevant. This method involves finding the *minimum spanning tree* of a distance or similarity matrix, and we therefore begin with a description of this.

Minimum spanning tree – technical details

Suppose n points are given (possibly in many dimensions), then a *tree* spanning these points, i.e. a *spanning tree*, is any set of straight-line segments joining pairs of points such that

1. No closed loops occur.
2. Each point is visited by at least one line.
3. The tree is connected, i.e. it has paths between any pair of points.

The length of a tree is defined to be the sum of the length of its segments, and when a set of n points and the lengths of all $\binom{n}{2}$ segments are given then we define the *minimum spanning tree* of the n points as the spanning tree of minimum length.

Algorithms to find the minimum spanning tree of a set of points given the distances between them are given in several papers, e.g. Prim (1957), Loberman and Weinberger (1957), and Gower and Ross (1969).

The links of the minimum spanning tree of the original distance matrix may be plotted onto the two-dimensional representation obtained from a particular ordination method, and the resulting diagram may, as the example in the next section illustrates, help to highlight distortions produced by the mapping process.

Using the minimum spanning tree to highlight 'distortions' produced by the ordination process

To illustrate the use of the minimum spanning tree for indicating 'inadequacies' in two-dimensional representations of multivariate data we shall examine some astronomical data given by Nathanson (1971). These data concern 273 galaxies each described by seven descriptive parameters such as diameter, surface brightness, colour, etc. The conventional morphological system of classification of galaxies (based essentially on their shapes) was used to classify these galaxies into ten groups, and then Mahalanobis distance (see Appendix A) between each pair of groups was calculated on the basis of the seven physical pro-

36

perties measured for each galaxy. The resulting distance matrix appears in Table 2.4. The application of principal co-ordinates analysis (see page 16) to this matrix results in the two-dimensional configuration shown in Figure 2.13. If we now find the minimum spanning tree of the distance matrix of Table 2.4 and plot its links onto Figure 2.13, we obtain Figure 2.14. The latter indicates that the mapping into two dimensions has produced some distortion of the original configuration of galaxies.

This is most easily seen if we consider the points in these diagrams representing the SBb and Sb groups of galaxies. The positions of these points apparently indicate that these two groups lie relatively close together and are well separated from the Sc group. The structure of the minimum spanning tree seen in Figure 2.14 shows, however, that the SBb and Sb groups are *closer* to the Sc group than to each other (otherwise the points representing them would be a line in a shorter tree), and that their apparent closeness in Figure 2.13 is illusory. Of course, that this is so may also be seen by examining the distances in Table 2.4. However, when n is large the distance matrix becomes too large to be assimilated and the minimum spanning tree is then a useful ancillary technique.

Table 2.4

Inter-group Mahalanobis distances for ten groups of galaxies (Taken from Nathanson (1971))

	1	2	3	4	5	6	7	8	9	10
1	0·00									
2	3·29	0·00								
3	2·79	1·13	0·00							
4	3·52	1·75	1·45	0·00						
5	3·77	2·97	1·71	2·02	0·00					
6	3·27	3·01	2·13	1·89	1·27	0·00				
7	3·93	3·72	3·00	2·25	1·86	0·68	0·00			
8	3·86	5·12	4·11	3·24	3·15	1·59	1·51	0·00		
9	3·77	5·70	4·85	3·85	3·41	1·74	2·05	0·91	0·00	
10	4·12	6·88	6·02	7·03	5·38	4·09	4·03	2·24	1·87	0·00

1 Irregulars (I)	2 SBc (SB1)	3 Sc (SC)	4 SBb (SB2)
5 Sb (SB)	6 SBa (SBA)	7 Sa (SA)	8 SBO (SBO)
	9 SO (SO)	10 Ellipticals (E)	

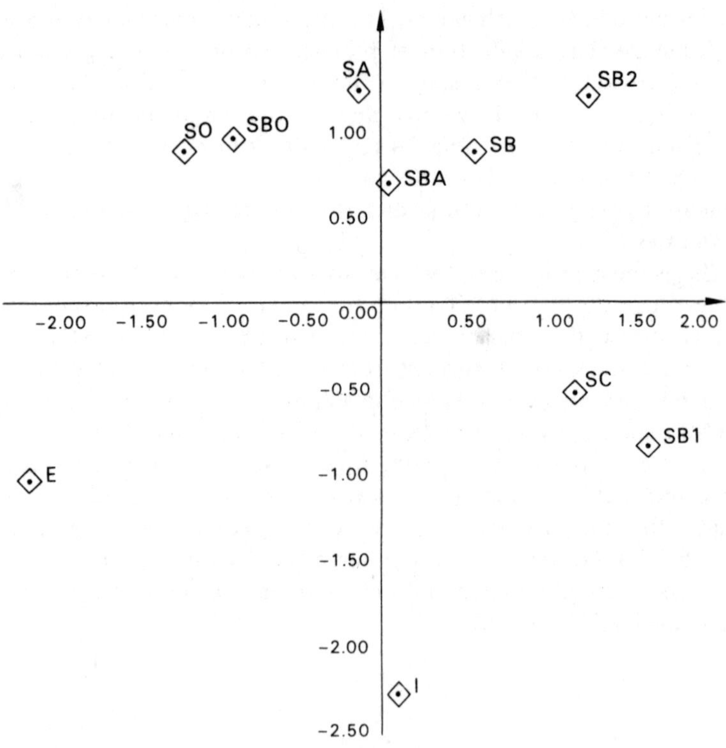

Figure 2.13   Two-dimensional solution given by principle co-ordinates analysis applied to the distance between galaxies matrix shown in Table 2.4.

A Comparison of Ordination Techniques

In this section we shall attempt, briefly, to indicate the similarities and differences between the various techniques discussed in this chapter and to point out some of their advantages and disadvantages.

The first point to make is that the first three techniques discussed, namely principal component plots, principal co-ordinate plots, and the biplot, are all latent root and vector procedures, whilst the remaining two techniques considered, multidimensional scaling and non-linear mapping, both operate by minimizing a particular function using some iterative algorithm. The former methods have, therefore, obvious computational advantages. Of these, principal co-ordinates is perhaps the most powerful for obtaining a low-dimensional representation of data

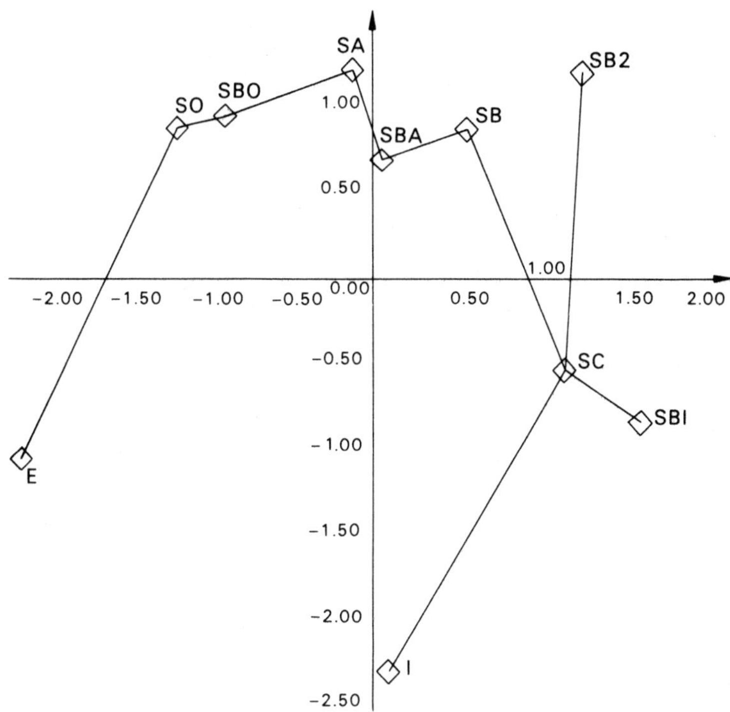

Figure 2.14   Minimum spanning tree of the distance between galaxies matrix (Table 2.4) plotted onto principal co-ordinates two-dimensional representation.

since it is not as restrictive as principal components analysis. In particular it is not necessary to consider only data sets for which Euclidean distance is considered appropriate. The only advantage of this particular distance measure is that it allows the principal co-ordinates to be related *linearly* to the original variable values. Principal co-ordinates analysis also has the advantage of being directly applicable to data *given* in the form of a distance or similarity matrix. The biplot obviously has considerable advantages if one wishes to obtain a simultaneous display of relationships between variables and between individuals. The particular form of this technique described on page 23, may almost be regarded as a combination of principal components and principal co-ordinates analysis.

In many respects the mathematical formulations of non-metric multidimensional scaling and non-linear mapping are similar. However, the mapping criteria, i.e. 'Stress' and 'Mapping error', are quite different. A major distinction is that multidimensional scaling employs only the ordinal properties of the similarities or distances being used, which in many cases may be thought a considerable advantage. Gower (1966) discusses the relationship between principal co-ordinates analysis and non-metric multidimensional scaling. He concludes that where the former gives an adequate fit in two dimensions, then the solution will be similar to the one that would be found employing the latter method. Gower points out, however, that multidimensional scaling may be able to find a good fit in a low number of dimensions when principal co-ordinates may not; because of the differing computational complexities of the two methods Gower finally recommends the initial use of principal co-ordinates analysis, and, where this does not lead to a solution of sufficiently low dimensionality, he suggests using the co-ordinates found as a starting point for the iterative algorithm of non-metric multidimensional scaling. Gnanadesikan and Wilk (1969) also point out that multidimensional scaling may produce suitable solutions for certain types of *curved* multidimensional data which would not be indicated by, say, principal components analysis. Sammon (1969) gives some interesting examples when non-linear mapping recovers the structure in some specially constructed sets of multivariate data, whilst principal component plots of the same data fail to indicate this correctly.

In general it would appear that the choice of a particular method from amongst those discussed in this chapter will be governed by many factors; e.g. the type of data to be analysed, the number of observations involved, the complexity of the structure expected, etc. No overall recommendation can be made as to the 'best' method to use in all possible situations; indeed it may not be unreasonable to use more than one of the techniques in many cases. For many sets of data the methods may all give very similar results. However, since in general fewer dimensions will be needed to reproduce ordinal information than to reproduce metric information, non-metric multidimensional scaling techniques are perhaps more *likely* to give useful and informative low-dimensional representations of data than other methods, especially when any structure is of a complex non-linear variety.

Summary

In this chapter several techniques which are useful for producing a low-dimensional representation of multivariate data have been discussed. The main interest here has been explicitly in the two-dimensional solution given by the methods, since our main aim has been to be able to examine the data visually. It should be mentioned, however, that for some data sets it would be unrealistic to expect a *two*-dimensional representation to give anything but a very approximate indication of the inherent structure present. In other words, two dimensions may just not be enough to accommodate the full complexity of the relations in the given data set. Unfortunately no real tests exist for the value of the number of dimensions necessary to provide an adequate fit, apart, of course, from the informal goodness of fit criteria already mentioned in connection with particular methods. However, Gnanadesikan and Wilk (1969) make the following important point which perhaps suggests that a formal test of number of dimensions is not of great importance.

'Interpretability and simplicity are important in data analysis and any rigid inference of optimal dimensionality, in the light of the observed values of a numerical index of goodness of fit, may not be productive.'

Two-dimensional solutions certainly have the virtue of simplicity; they are also readily assimilated by the investigator and may, in many cases, provide an easily understood basis for the discussion of complex multivariate data; consequently they are likely to be the solutions of most *practical* importance.

Finally it should be mentioned again that the techniques described in this chapter should in no way be regarded as methods to be used to the *exclusion* of other more familiar types of multivariate analysis; indeed they will, in general, be *most* helpful when used alongside, and in association with, other forms of analysis.

3. The Graphical Display of Clusters, and Similarity Matrices

Introduction

In the previous chapter several methods for obtaining a low-dimensional representation of multivariate data were described. Some of these, e.g. principal co-ordinates analysis, could be usefully applied to a similarity or distance matrix to obtain a geometrical representation of the relationships between observations implied by the elements of such a matrix. In this chapter some further methods which may be helpful in analysing similarity and distance matrices and which again result in some form of graphical or visual output are described. Perhaps the most well known of these are those clustering techniques which when applied to this type of matrix lead to a *dendrogram*, and we begin with a description of some of these.

Hierarchical Clustering Techniques – Dendrograms

Clustering techniques seek to form 'clusters', 'groups', or 'classes' of individuals, such that individuals within a cluster are more 'similar' in some sense than individuals from different clusters. Comprehensive reviews of the many available methods are given by Cormack (1971) and Everitt (1974). Here we shall concentrate on a particular set of techniques which produce a solution in which some clusters are nested within other clusters; these are known as *hierarchical techniques*, and may be subdivided into *agglomerative* methods which proceed by a series of successive fusions of the *n* individuals into groups, and *divisive* methods which partition the set of individuals successively into finer groupings. Both agglomerative and divisive techniques operate on a matrix of inter-individual similarities or distances, and the results they give may be presented in the form of a *dendrogram*, this being a two-dimensional diagram illustrating the fusions (agglomerative methods) or divisions (divisive methods) which have been made at each stage of the procedure. (Examples of dendrograms will be given later.) These methods are well described in the reviews previously mentioned, and here we give only a brief description of agglomerative hierarchical techniques, and then indicate the possible advantages and disadvantages

42

of using the dendrograms produced as a method for representing similarity and distance matrices.

Agglomerative hierarchical techniques – technical details
The basic procedure with all these methods is similar. Beginning with the inter-individual similarity or distance matrix the methods fuse individuals or groups of individuals which are closest (or most similar), and thereby proceed from the initial stage of n individuals to the final stage in which all individuals are in a single group. The complete procedure is described by the dendrogram. Differences between methods arise because of the different ways of defining distance (or similarity) between an individual and a group containing several individuals, or between two groups of individuals. The inter-group distance measures used by the majority of these techniques satisfy a recurrence formula for the distance, $d_{k(ij)}$, between a group k and a group (ij) formed by the fusion of groups i and j. This formula is

$$d_{k(ij)} = \alpha_i d_{ki} + \alpha_j d_{kj} + \beta d_{ij} + \gamma |d_{ki} - d_{kj}| \qquad \textbf{3.1}$$

where d_{ij} is the distance between groups i and j, and α, β, and γ are parameters whose values for the most commonly used agglomerative techniques are given in Table 3.1.

Table 3.1

Recurrence formula parameters for various agglomerative hierarchical clustering methods

1. Nearest neighbour – single linkage: $\alpha_i = \alpha_j = \frac{1}{2}; \ \beta = 0; \ \gamma = -\frac{1}{2}.$

2. Furthest neighbour – complete linkage: $\alpha_i = \alpha_j = \frac{1}{2}; \ \beta = 0; \ \gamma = \frac{1}{2}.$

3. Centroid clustering: $\alpha_i = n_i/(n_i + n_j); \ \alpha_j = n_j/(n_i + n_j); \ \beta = -\alpha_i \alpha_j;$
$$\gamma = 0.$$

4. Median clustering: $\alpha_i = \alpha_j = \frac{1}{2}; \ \beta = -\frac{1}{4}; \ \gamma = 0.$

5. Group average: $\alpha_i = n_i/(n_i + n_j); \ \alpha_j = n_j/(n_i + n_j); \ \beta = \gamma = 0.$

6. Ward's method: $\alpha_i = \dfrac{n_k + n_i}{n_k + n_i + n_j}; \ \alpha_j = \dfrac{n_k + n_j}{n_k + n_i + n_j};$

$$\beta = \dfrac{-n_k}{n_k + n_i + n_j}; \ \gamma = 0.$$

At this point it is convenient to give an example of a dendrogram for readers unfamiliar with the concept. For this purpose, single linkage clustering was applied to the following matrix of distances between five individuals.

$$
D = \begin{array}{c@{\quad}c}
 & \begin{array}{ccccc} 1 & 2 & 3 & 4 & 5 \end{array} \\
\begin{array}{c} 1 \\ 2 \\ 3 \\ 4 \\ 5 \end{array} &
\left[\begin{array}{ccccc}
0{\cdot}0 & 2{\cdot}0 & 6{\cdot}0 & 10{\cdot}0 & 9{\cdot}0 \\
2{\cdot}0 & 0{\cdot}0 & 5{\cdot}0 & 9{\cdot}0 & 8{\cdot}0 \\
6{\cdot}0 & 5{\cdot}0 & 0{\cdot}0 & 4{\cdot}0 & 5{\cdot}0 \\
10.0 & 9{\cdot}0 & 4{\cdot}0 & 0{\cdot}0 & 3{\cdot}0 \\
9{\cdot}0 & 8{\cdot}0 & 5{\cdot}0 & 3{\cdot}0 & 0{\cdot}0
\end{array} \right]
\end{array}
$$

The resulting dendrogram is shown in Figure 3.1, and it is this that now acts as our 'visual' representation of D. (For the details of how single linkage clustering operates on D to produce Figure 3.1, see Everitt, 1974, Chapter Two). The dendrogram constitutes a *summary* of the information in D, since in it many of the original inter-individual distances are lost. It therefore becomes of some importance to seek to measure the goodness of fit of the dendrogram for the distance matrix,

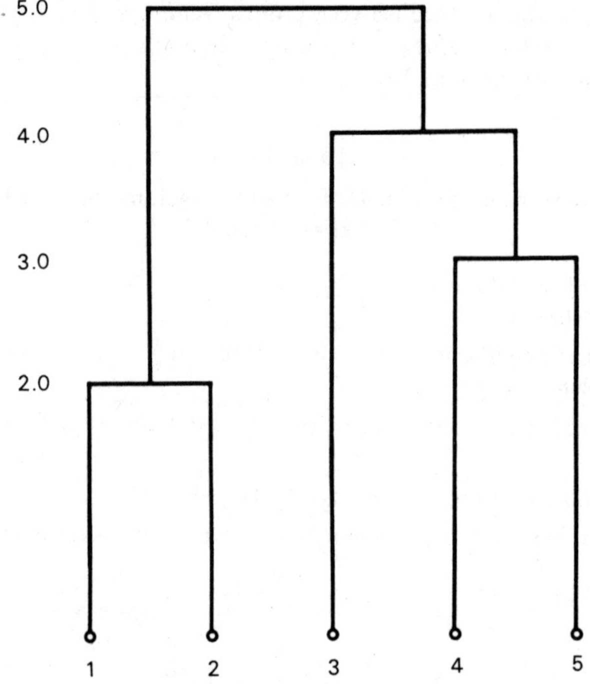

Figure 3.1   Single linkage dendrogram for the distance matrix, D.

D. As mentioned in the previous chapter in respect of ordination techniques, goodness of fit in this context is judged by how well the original inter-individual distances are preserved by the representation involved, in this case the dendrogram. It therefore becomes necessary to decide the inter-individual distances implied by the dendrogram, so that these may then be compared with the original distances in some way (see page 35). Hartigan (1967) and Jardine *et al.* (1967) show that the distance between two individuals in a dendrogram may be taken as the fusion level at which they both appear, for the first time, in the same cluster.

For example, for the dendrogram shown in Figure 3.1 the distances between the five individuals would be taken as follows:

$$
D_{(\text{dendrogram})} = \begin{array}{c} \\ 1 \\ 2 \\ 3 \\ 4 \\ 5 \end{array}
\begin{array}{ccccc}
1 & 2 & 3 & 4 & 5 \\
\left[\begin{array}{ccccc}
0{\cdot}0 & 2{\cdot}0 & 5{\cdot}0 & 5{\cdot}0 & 5{\cdot}0 \\
2{\cdot}0 & 0{\cdot}0 & 5{\cdot}0 & 5{\cdot}0 & 5{\cdot}0 \\
5{\cdot}0 & 5{\cdot}0 & 0{\cdot}0 & 4{\cdot}0 & 4{\cdot}0 \\
5{\cdot}0 & 5{\cdot}0 & 4{\cdot}0 & 0{\cdot}0 & 3{\cdot}0 \\
5{\cdot}0 & 5{\cdot}0 & 4{\cdot}0 & 3{\cdot}0 & 0{\cdot}0
\end{array}\right]
\end{array}
$$

Such distances satisfy the *ultrametric inequality*, namely that for a distance measure d_{ij}

$$d_{ij} \leqslant \max\{d_{ik}, d_{jk}\} \qquad\qquad \textbf{3.2}$$

The goodness of fit of the dendrogram may now be assessed by comparing the original distances with these ultrametric distances. Various statistics have been suggested for making such a comparison. One, which was briefly mentioned in the previous chapter (see page 35), is the cophenetic correlation coefficient. Another, again mentioned in Chapter 2, is δ, given by

$$\delta = \sum_{i<j} |d_{ij} - d_{ij}^{*}| \qquad\qquad \textbf{3.3}$$

where the d_{ij} are the original distances, and the d_{ij}^{*} are the ultrametric distances given by the dendrogram. Jardine and Sibson (1968) propose a series of measures for comparing the d_{ij} and the d_{ij}^{*}; these are given by the following general formula:

$$\delta_{\mu} = \left[\sum_{i<j} |d_{ij} - d_{ij}^{*}|^{1/\mu}\right]^{\mu} \Big/ \left[\sum d_{ij}^{*\,1/\mu}\right]^{\mu}, \quad 0 \leqslant \mu \leqslant 1 \qquad \textbf{3.4}$$

These measures are essentially standardized Minkowski metrics, and by varying μ greater or lesser emphasis is placed on larger or smaller differences. A perfect fit would be indicated by a zero value of δ_μ. A comparative sampling study of several of these goodness of fit measures is reported by Gower and Banfield (1974). For the dendrogram of Figure 3.1, the values taken by some of the measures are as follows:

1. Cophenetic correlation: $r = 0.82$.

2. $\delta_{1/2} = \left[\sum (d_{ij} - d_{ij}^*)^2 \right]^{1/2} \bigg/ \left[\sum d_{ij}^2 \right]^{1/2} = 0.39$

3. $\delta_1 = \sum |d_{ij} - d_{ij}^*| \bigg/ \sum d_{ij} = 0.29$.

A dendrogram representation of a similarity or distance matrix is likely to be most useful if the data are strongly clustered *and* hierarchical. In other cases such a representation may be very misleading (see, for example, some of the dendrograms in Everitt (1974, Chapter 4)). Rohlf (1970) suggests that when comparing different sets of data using the same clustering procedure, one may use the measures of goodness of fit already discussed as indices to measure the extent to which the cluster structure is hierarchic. If, for example, the cophenetic correlation is very high, say above 0.95, then one may safely conclude that imposing a system of nested clusters on the data has not caused undue distortion. If the correlation is much lower, say about 0.6 or 0.7, then the assumption that the data consist of a system of nested clusters is questionable.

A further problem arises in attempting to display the dendrogram on the printed page. In order to do this some linear ordering of the individuals must be selected. However, any selected ordering may be misleading in that adjacent individuals may be quite dissimilar, and individuals in widely separated stems of the dendrogram may be more similar to each other than to some of the individuals in their own stems.

To assist with understanding the possible distortions produced by the dendrogram, Rholf (1970) suggests plotting frequency distributions of the inter-individual distance measures on each of the branches of the dendrogram. Since each branch unites two clusters, the coefficients in the frequency distributions are the distances between all the individuals in one branch and the individuals in another. A detailed study of such frequency distributions may be useful in indicating the 'fine' structure of the clusters implied by the dendrogram.

Let us now examine in more detail some dendrogram representations of various distance and similarity matrices.

Table 3.2

Distance matrix between twenty individuals

	1	2	3	4	5	6	7	8	9	10	11	12	13	14	15	16	17	18	19	20
1	0·0																			
2	0·539	0·0																		
3	1·438	0·780	0·0																	
4	0·462	0·006	0·713	0·0																
5	0·311	0·312	0·413	0·236	0·0															
6	12·130	17·306	21·297	17·008	15·988	0·0														
7	3·813	5·830	9·834	5·795	6·266	4·672	0·0													
8	4·063	7·536	9·210	7·257	6·055	2·811	2·992	0·0												
9	5·568	9·555	11·217	9·233	7·783	2·219	3·922	0·125	0·0											
10	3·619	6·695	9·084	6·573	5·772	2·565	1·722	0·178	0·473	0·0										
11	39·890	42·383	32·645	41·580	35·994	61·704	61·666	41·778	41·864	46·429	0·0									
12	42·681	46·775	37·402	45·859	39·675	57·290	61·902	40·152	39·482	45·166	1·202	0·0								
13	32·237	35·506	27·277	34·719	29·409	48·338	50·241	31·599	31·377	35·866	0·935	0·811	0·0							
14	34·448	37·838	29·317	37·025	31·534	50·536	52·866	33·567	33·274	37·996	0·778	0·510	0·038	0·0						
15	34·203	38·176	30·034	37·335	31·730	47·257	51·222	31·642	31·085	36·100	1·681	0·508	0·231	0·190	0·0					
16	32·290	38·671	33·089	37·766	32·115	31·157	41·641	22·505	20·901	26·665	11·505	6·476	6·209	6·351	4·392	0·0				
17	31·646	38·080	32·709	37·182	31·596	29·940	40·533	21·656	20·036	25·744	12·208	7·074	6·684	6·861	4·831	0·015	0·0			
18	28·353	33·975	28·328	33·130	27·809	30·454	38·710	20·630	19·389	24·572	9·080	5·177	4·314	4·545	2·992	0·268	0·318	0·0		
19	32·302	37·508	30·643	36·633	30·990	38·148	45·459	26·052	24·958	30·372	5·558	2·415	2·170	2·216	1·127	1·071	1·297	0·525	0·0	
20	37·960	45·455	40·207	44·476	38·442	31·677	45·381	25·076	22·950	29·456	16·626	9·914	10·334	10·392	7·786	0·592	0·524	1·642	3·034	0·0

A dendrogram representation of a distance matrix
The distance matrix shown in Table 3.2 arises from finding the Euclidean distance between twenty individuals each having scores on two variables. The dendrogram produced by applying group average clustering to this matrix is shown in Figure 3.2. These data were generated to contain four well-separated and hierarchically structured clusters, and this structure is clearly reflected in the dendrogram. The cophenetic correlation coefficient takes the value 0·91, again clearly indicating the hierarchical nature of these clusters.

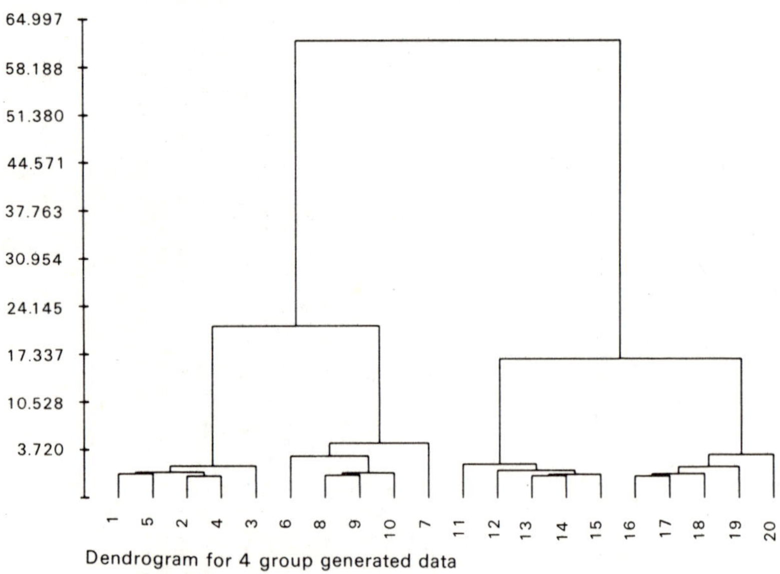

Figure 3.2   Complete linkage dendrogram for the distance matrix shown in Table 3.2.

A dendrogram representation of a similarity matrix
As a further example of the use of dendrogram representations, complete linkage clustering was applied to the correlation matrix shown in Table 3.3. This matrix arises from the studies reported by Jasper (1949) into job aptitudes, and has been the subject of various factor analytic investigations. The resulting dendrogram appears in Figure 3.3 and the cophenetic correlation coefficient takes the value 0·74; such a value would suggest that the assumption of a hierarchical structure for these correlations is perhaps questionable. In such cases it is often useful to obtain some further visual representation of the similarity matrix

48

Table 3.3

Correlation matrix from Jasper (1949), $N = 594$ (all decimal points omitted)

	V1	V2	V3	V4	V5	V6	V7	V8	V9	V10	V11	V12	V13	V14	V15	V16	V17	V18	V19	V20	V21	V22	V23
V2	560																						
V3	416	419																					
V4	282	238	407																				
V5	185	199	249	234																			
V6	366	300	537	557	324																		
V7	324	294	631	421	330	506																	
V8	369	328	585	500	251	576	442																
V9	400	273	506	407	276	452	581	374															
V10	376	272	463	403	275	465	492	377	754														
V11	353	307	619	417	328	551	766	498	600	555													
V12	282	241	463	421	346	523	664	360	572	521	707												
V13	312	276	520	431	332	471	703	318	622	591	684	725											
V14	392	353	598	348	177	420	515	481	414	340	491	364	426										
V15	634	599	559	273	240	399	427	424	407	374	468	334	375	479									
V16	448	489	420	237	217	319	342	392	299	270	322	248	262	362	558								
V17	353	296	548	499	269	541	407	563	321	306	426	360	353	402	372	303							
V18	304	190	411	478	279	533	468	392	535	541	536	565	525	291	261	194	368						
V19	276	240	408	548	271	544	443	437	363	397	423	435	413	349	277	265	453	418					
V20	349	310	349	222	126	206	236	247	181	130	184	114	164	304	389	312	227	133	223				
V21	386	378	410	278	147	253	337	277	282	231	263	217	294	402	470	311	254	182	245	701			
V22	243	238	244	247	144	240	257	275	224	215	214	161	209	282	315	276	165	203	233	397	452		
V23	345	316	354	264	139	288	289	310	264	218	300	215	241	302	388	301	261	181	306	561	580	511	
V24	368	329	338	241	110	241	263	309	221	193	241	179	212	384	406	337	239	152	215	418	448	447	431

so that the structure implied by the dendrogram may be examined further. Consequently Kruskal's method of non-metric multidimensional scaling was applied to the correlation matrix of Table 3.3 and the resulting two-dimensional configuration is shown in Figure 3.4. The stress value for this configuration was 30 per cent, indicating a fairly poor fit; nevertheless by embedding the hierarchical clustering solution in this two-dimensional multidimensional scaling solution (as shown in Figure 3.4), we obtain a fairly clear idea of the relationships between the variables implied by the correlations of Table 3.3. Clearly

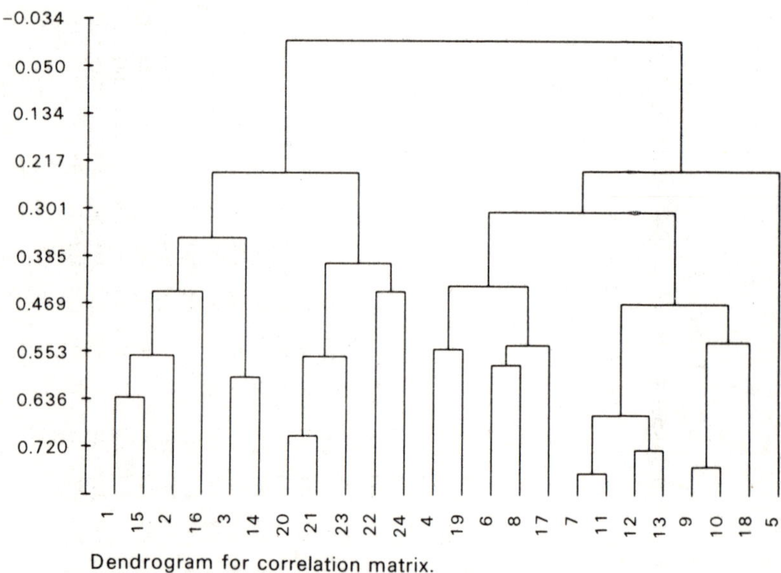

Figure 3.3　Complete linkage dendrogram for the correlation matrix shown in Table 3.3.

variable five is unrelated to the remainder. Variables one, two, fifteen, and sixteen form a fairly clear-cut cluster, as do variables twenty to twenty-four. The two-dimensional configuration shows that the two dendrogram clusters – variables four, six, eight, seventeen, and nineteen and variables seven, nine, ten, eleven, twelve, thirteen, and eighteen – are not very clearly separated. The dendrogram of Figure 3.3 and the two-dimensional diagram of Figure 3.4 lead, for these data, to very similar conclusions as those obtained from a factor analysis. However, researchers might find diagrams of this type an aid in the interpretation of the usual factor analysis results; indeed, many may find such diagrams a desirable alternative.

50

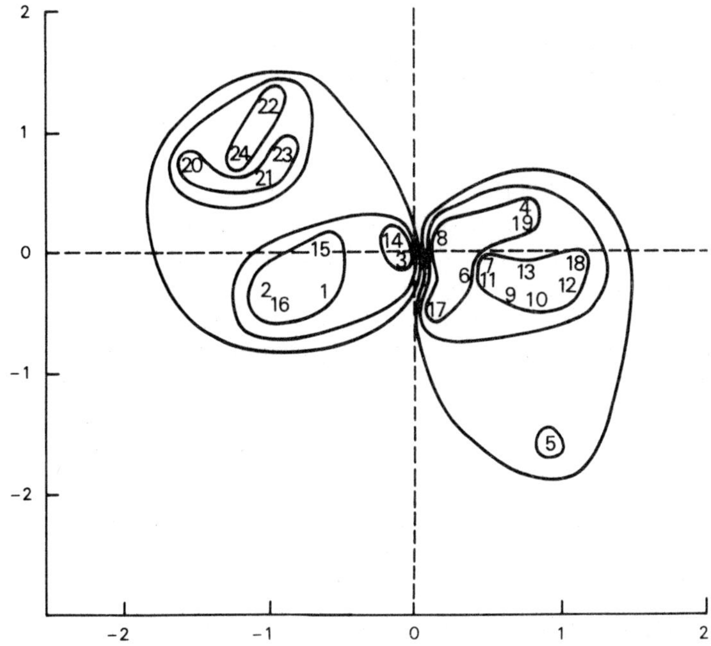

Figure 3.4   Non-metric multidimensional scaling configuration for the correlation matrix in Table 3.3, with imbedded complete linkage clustering solution.

Direct Graphical Displays of Similarity and Distance Matrices

Various authors have suggested a simple method for obtaining a visual display of a similarity or distance matrix which involves representing the elements of the matrix by symbols having varying characteristics, these depending on the value of the particular similarity or distance involved. For example, Carmichael and Sneath (1969) use dots of various sizes, the larger ones representing higher similarity values. Other authors suggest the use of symbols of different colours. Such techniques may be especially helpful if the order of the individuals in the matrix is changed so that the highest similarities (smallest distances) are arranged closest to the diagonal. When there are more than a few individuals in the data this reordering may be difficult, since there may be no single ordering of the individuals which brings all the higher similarities close to the diagonal. Ling (1973) overcomes this problem using an ordering of individuals arising from some form of cluster analysis, by rearranging the original sequence of individuals so that members of the same cluster always lie in consecutive rows and columns of the permuted matrix. This procedure which Ling calls SHADE is now described in more detail.

SHADE – technical details

The principle used in this technique is to represent the elements of the similarity or distance matrix by characters of varying shades of darkness, high similarities (low distances) being indicated by darker characters. In the computer program produced by Ling for this technique (see Appendix B) the symbols used to represent the similarities or distances are produced by character overprinting. Fifteen symbols (shown in Figure 3.5 in increasing degree of darkness) are used to

Symbols	.	•	▾	=	+	⬧	✳	X	I	⬧	▮	▨	▦	▨	▮
class	1	2	3	4	5	6	7	8	9	10	11	12	13	14	15

Figure 3.5   SHADE procedure symbols.

represent the similarities or distances after the latter have been grouped into one of fifteen classes. The following four methods of defining the classes are available:

1. The ranks of the $\binom{n}{2}$ distances are divided into fifteen equal intervals and the i-th character printed for distance d_{kl} if

$$(15-i)\binom{n}{2}\Big/ 15 < \text{rank of } d_{kl} \leqslant (16-i)\binom{n}{2}\Big/15 \qquad (i=1, 2, \ldots, 15) \qquad \textbf{3.5}$$

2. The actual distances are used and divided into fifteen equal intervals between the minimum (d_{min}) and maximum (d_{max}) values amongst the $\binom{n}{2}$ distances, and the i-th character is printed for distance d_{kl} if

$$d_{min} + (15-i)\delta < d_{kl} \leqslant d_{min} + (16-i)\delta \qquad (i=1, 2\ldots, 15) \qquad \textbf{3.6}$$

where $\delta = (d_{max} - d_{min})/15$.

3. The user specifies d_{min} and d_{max}, the minimum and maximum distance values for defining the class intervals, and then the i-th character is plotted for distance d_{kl} if equation 3.6 applies. To ensure that this option is applicable to all the distances we let d_{kl} be included in the first class if $d_{kl} > d_{max}$ and in the fifteenth class if $d_{kl} < d_{min}$.

4. All class intervals $c_0, c_1, \ldots, c_{15}$ are defined by the user and the i-th symbol is printed for d_{kl} if

$$c_{15-i} < d_{kl} \leqslant c_{16-i} \qquad \textbf{3.7}$$

The case of $d_{kl} < c_0$ or $d_{kl} > c_{15}$ are handled as for $d_{kl} < d_{min}$ and $d_{kl} > d_{max}$ in option 3.

The above methods of defining classes are described in terms of distances. If we have a matrix of similarities then the options are defined analogously, with the darkest symbols corresponding to the highest similarities instead of the smallest distances. This would mean that the symbols in Figure 3.5 would now be renumbered 15, 14, ..., 3, 2, 1. In either case the diagonal elements are always represented by the darkest symbol.

The first of the four options described above uses only the ranks of the distance or similarity values, whilst the other three use the actual numerical values of these measures. The choice of which particular option to use for any particular similarity or distance matrix depends on several factors. Firstly if one wishes to use only the ordinal properties of the coefficients then option 1 *must* be used. If there are one or two very large values in the matrix relative to the others then option 2 is unlikely to be very useful, since the partition of the distances or similarities into fifteen equally spaced classes between the maximum and minimum values of the coefficients may leave all but the extreme classes empty. The resulting plot would therefore consist of only two symbols and, consequently, would not be very informative. This problem may be overcome by using the third option, and choosing a value of d_{max} larger than all but the few extremely large coefficient values in the matrix.

Using one of the four options described the procedure is applied to the similarity or distance matrix before clustering and to the rearranged matrix after clustering (i.e. with the individuals now in 'cluster order'). The presence of well-separated clusters in the data would now be indicated by clearly defined dark triangles immediately under the diagonal of the rearranged matrix, as we shall illustrate in the following examples.

SHADE applied to a distance matrix

As the first example of the SHADE procedure it was applied to a matrix of Euclidean distances obtained from thirty individuals each having measurements on five variables. These data had been artificially generated to contain three fairly well separated clusters. The results obtained are shown in Figures 3.6 and 3.7. The first of these corresponds to the distance matrix with the individuals arranged in a 'random order', and the second to the distance matrix rearranged so that individuals are in their 'cluster order'. In both cases the second

Generated data in random order

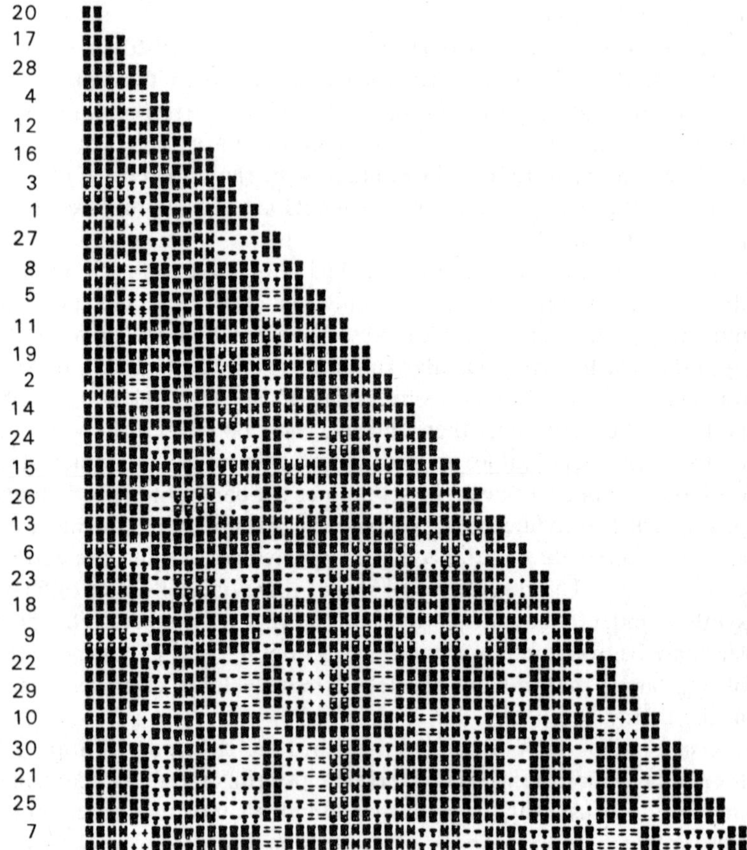

Figure 3.6   SHADE example, individuals in 'random order'.

of the four printing options described in the previous section was used.

Figure 3.7 clearly shows the presence of dark triangles immediately below the diagonal indicating quite dramatically the structure of the data as implied by the distance coefficients.

A further example of the use of SHADE

As a further example of the operation of the SHADE procedure it was applied to the matrix of Mahalanobis distances between twenty-two Indian tribal populations given by Mahalanobis, Majumdar, and

54

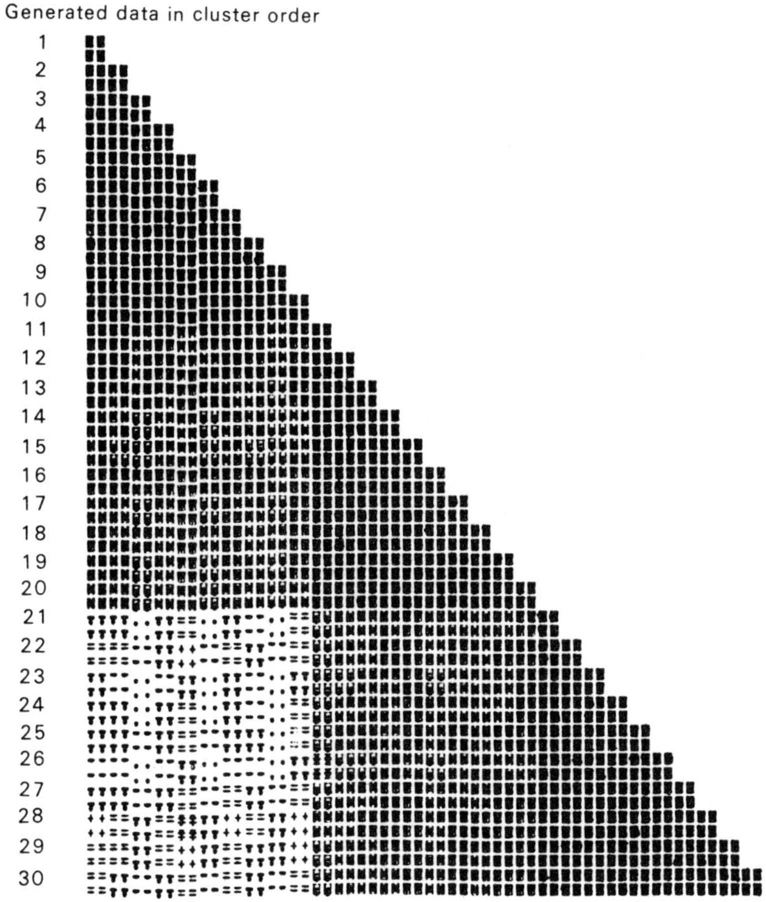

Generated data in cluster order

Figure 3.7   SHADE example, individuals in 'cluster order'.

Rao (1949) and shown here in Table 3.4. This matrix arises from the measurement of twelve variables in samples of about 160 from each of the twenty-two tribes. Two representations of the matrix of Table 3.4 given by thc SHADE procedure are shown in Figures 3.8 and 3.9. The first of these arises when the matrix is arranged so that the twenty-two tribes are in a random order, and the second when the tribes are arranged in order corresponding to the 'clusters' suggested by Mahalanobis. (*See* Mahalanobis *et al*, op. cit., part II.) In Figure 3.9 three fairly clear dark triangles are visible. The first of these corresponds to the two Brahmin tribes (tribes one and two). The second

55

Table 3.4

Mahalanobis distance between twenty-two Indian tribes (taken from Mahalanobis *et al.* (1949)).

	B_1	B_2	Ag	Ch	M	C_1	C_2	Bh	D	A_1	A_2	A_3	A_4	Th	Cm	T_1	T_2	T_3	T_4	T_5	T_6	T_7
B_1	0·0																					
B_2	0·27	0·0																				
Ag	3·17	1·94	0·0																			
Ch	2·98	2·81	2·33	0·0																		
M	2·83	2·61	2·18	0·39	0·0																	
C_1	1·53	1·63	3·50	2·50	2·07	0·0																
C_2	1·32	0·69	2·00	3·23	2·53	1·12	0·0															
Bh	4·41	3·79	3·46	5·01	3·15	2·56	2·17	0·0														
D	2·82	2·78	3·21	3·84	2·47	1·98	1·95	1·15	0·0													
A_1	1·13	0·73	2·58	3·18	2·33	1·26	0·71	2·38	2·75	0·0												
A_2	1·44	0·98	1·95	2·11	1·21	1·56	0·98	2·07	2·34	0·30	0·0											
A_3	2·11	1·45	2·07	2·58	1·37	1·73	0·99	1·65	2·20	0·48	0·11	0·0										
A_4	3·26	2·68	2·40	2·03	0·78	2·77	2·31	2·09	2·50	1·52	0·58	0·43	0·0									
Th	6·51	5·04	4·16	5·83	4·23	5·50	3·51	3·90	5·06	4·49	3·26	2·65	3·24	0·0								
Cm	5·40	4·32	2·12	2·30	1·23	3·75	3·22	1·91	2·90	2·94	1·69	1·28	0·59	2·75	0·0							
T_1	10·21	7·88	4·79	5·75	3·58	7·66	6·48	3·10	4·35	6·45	3·65	3·58	2·31	3·02	1·36	0·0						
T_2	9·10	8·77	5·29	7·39	4·88	8·83	7·31	3·58	4·97	7·26	5·16	4·25	3·09	3·42	1·99	0·29	0·0					
T_3	9·63	8·37	5·41	5·87	3·61	7·60	6·85	3·49	5·29	6·20	4·17	3·35	2·02	3·42	1·15	0·37	0·37	0·0				
T_4	7·97	6·72	4·39	5·46	3·42	6·22	5·22	2·50	4·41	5·11	3·35	2·67	2·04	1·79	1·33	0·48	0·77	0·56	0·0			
T_5	10·41	8·23	4·87	7·50	5·88	9·93	7·10	6·30	8·78	7·16	5·22	4·53	4·17	1·72	3·25	2·48	2·57	2·53	1·60	0·0		
T_6	7·69	5·71	3·53	5·36	4·31	6·49	4·47	4·75	7·71	4·16	3·12	2·67	2·93	2·01	2·39	3·89	4·47	3·36	2·31	1·20	0·0	
T_7	10·39	8·30	8·30	9·91	7·01	10·18	7·79	5·56	7·85	7·20	5·43	4·56	4·21	2·29	4·50	2·41	2·43	2·44	1·64	2·13	3·92	0·0

Indian dist matrix in random order

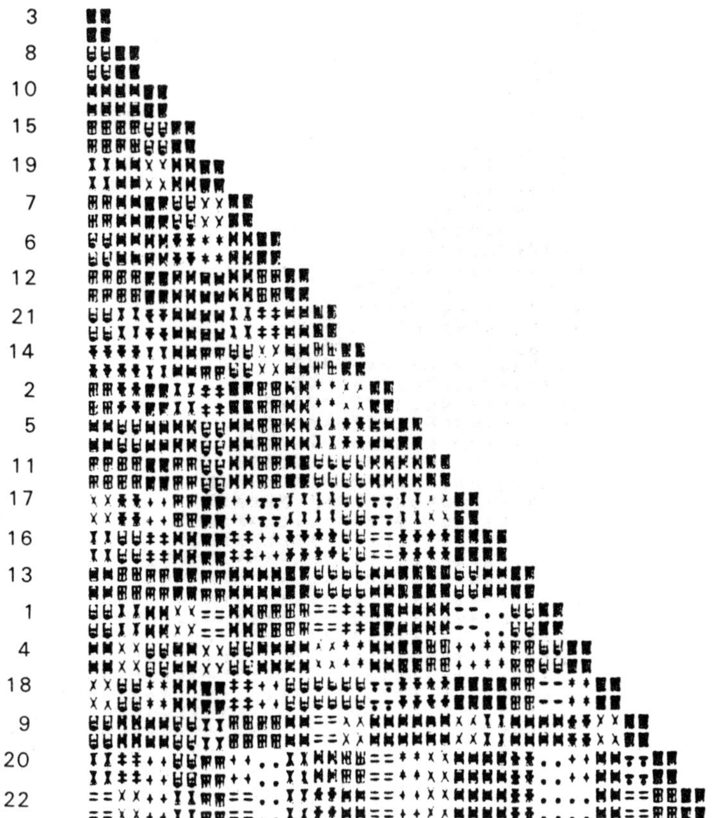

Figure 3.8   SHADE example, tribes in 'random order'.

corresponding to tribes ten, eleven, twelve, and thirteen is, essentially, the Artisan cluster mentioned by Mahalanobis. The third dark triangle corresponds to tribes sixteen, seventeen, eighteen, and nineteen. These are a subset of the third major cluster found in the original analysis. Figure 3.9 also indicates that the remaining tribes are fairly distinct from one another. It is also clear from this diagram that of the three major clusters of tribes identified, cluster one (i.e. tribes one and two) and cluster three (i.e. tribes sixteen to nineteen) are at some distance from each other, and that cluster two lies somewhere between them, although possibly marginally closer to cluster one.

57

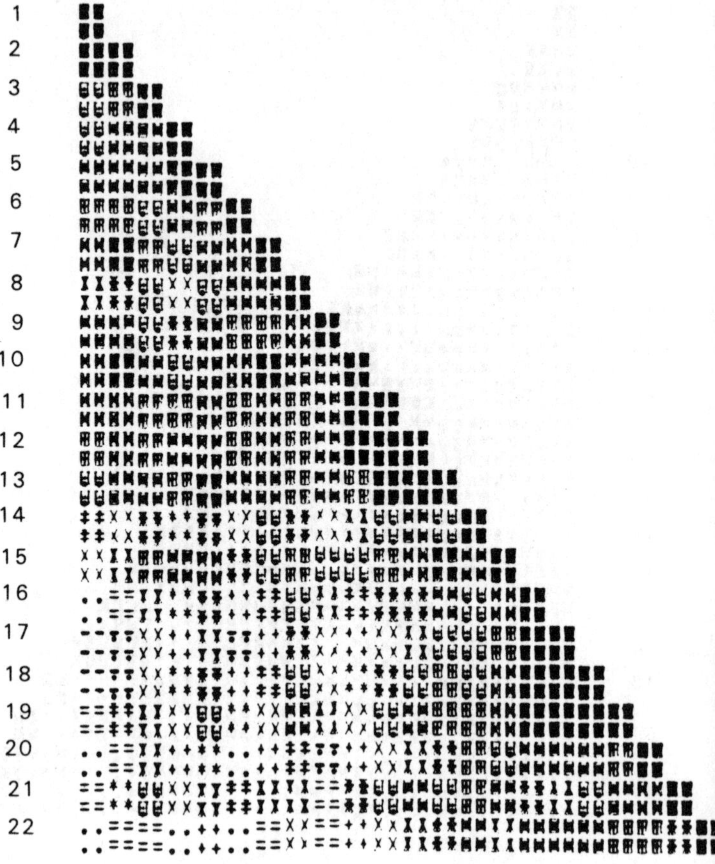

Figure 3.9   SHADE example, tribes in 'clusters order'.

Some Further Methods for Displaying Clustering Solutions

The methods described in the previous two sections have been discussed
in terms of representing similarity or distance matrices. They might
equally well have been regarded as ways of displaying the results of a
cluster analysis applied to such a matrix. In this section some further
methods of displaying clustering solutions will be considered. These
techniques are of importance because they give some indication of the
'reality' of the 'clusters' which are, in general, the inevitable result of

58

applying cluster analysis to data, no matter what the actual structure present. (Of course, many of the ordination techniques described in Chapter 2 would also be suitable for this purpose.)

Cannonical variates

A familiar technique of many multivariate investigations is *discriminant function analysis*, first proposed by Fisher (1936). This technique seeks to find a linear function of the variables which best discriminates between two different groups of individuals, these groups having been defined *a priori* by some criterion external to the set of variables measured. The function found may be used in problems of the identification and classification of 'new' individuals. The extension of this technique to a situation involving more than two groups is known as *cannonical variate analysis*. (*See*, for example, Maxwell (1961).) In this case several orthogonal linear functions of the original variables are determined. The technique may be used to display the results of a cluster analysis by plotting an individual's cannonical variate scores, and this is often useful for indicating how well separated or otherwise are the clusters found. Let us now look in more detail at this type of analysis.

Cannonical variates – technical details

We assume that the members of a population may be classified into k exhaustive and mutually exclusive groups or categories, and that a sample from each of these groups is available. For each individual, measurements of p variables are obtained. The problem is to find a linear function, or orthogonal linear functions, which will best discriminate between the groups. The functions will be of the form

$$f(\mathbf{x}) = w_1 x_1 + w_2 x_2 + \ldots + w_p x_p \qquad \textbf{3.8}$$

where the w's are weights which have to be determined.

The problem may be formulated as one of finding that set (or sets) of weights which maximizes the expression

$$\lambda = (\mathbf{w}'\mathbf{B}\mathbf{w})/(\mathbf{w}'\mathbf{W}\mathbf{w}) \qquad \textbf{3.9}$$

where $\mathbf{B}$ is the 'between groups' $(p \times p)$ matrix of sums of squares and cross-products of the variables about their respective means, and $\mathbf{W}$ is the corresponding 'within groups' matrix. It is easy to show that choosing $\mathbf{w}$ to maximize λ is equivalent to maximizing the differences between group means on the composite weighted x-scores (i.e. $f(\mathbf{x})$ of equation 3.8).

On differentiating 3.9 with respect to the **w**'s and equating to zero we obtain the equation

$$\mathbf{w}'(\mathbf{BW}^{-1} - \lambda\mathbf{I}) = 0 \qquad\qquad \textbf{3.10}$$

The solution of 3.10 involves finding the latent roots and vectors of the assymmetric matrix $\mathbf{BW}^{-1}$. The number of latent roots possible will depend on the rank of this matrix, which can be shown to be equal to p or $(k-1)$, whichever is the smaller. It follows that when only two groups are concerned only one set of weights and, consequently, only one cannonical variate results. This will be the linear discriminant function of Fisher.

The cannonical variate scores for each individual may now be found using equations of the form of 3.8, and then pairs of these may be plotted. Since these variates account for decreasing portions of the between groups variance, plots involving the first two or three are generally the most helpful.

An example of using cannonical variate plots to display the results of a cluster analysis
A set of data consisting of eight scores for each of eighty-six long-term prisoners in a London prison were subjected to cluster analysis using the technique described by Wolfe (1970) involving mixtures of multivariate normal distributions. A three-cluster solution was produced and this was then subjected to a cannonical variates analysis. Since there are only three groups, only two such variates arise, and a plot of these data in the space of these two variates is shown in Figure 3.10. The cannonical variate weights are shown in Table 3.5. The major impression gained from this plot is of a 'cloud' of points with little evidence of any clear-cut 'gaps' between the clusters. The diagram would suggest that this cluster solution has not identified really distinct groups of prisoners, but has merely 'dissected' a homogeneous data set into three parts.

Taxometric maps
Several authors, e.g. De Ley (1962) and Moss (1967), have presented methods of displaying classifications in the form of a diagram with circles or boxes representing classes, and lines between them representing their relationship to one another. Hence we shall describe a method of this kind suggested by Carmichael and Sneath (1969), who name the displays *taxometric maps*.

60

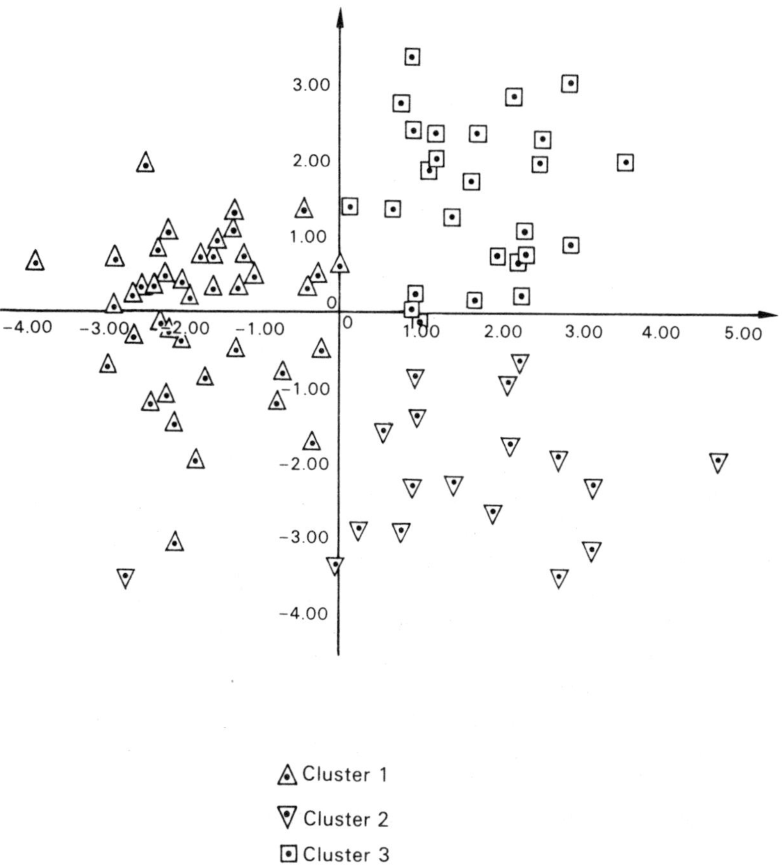

Figure 3.10 Cannonical variates plot for prisoners' data.

Taxometric maps – technical details

In a taxometric map diagram to display a classification, clusters are
represented by circles whose diameter is proportional to the distance
between the two most distant points in the cluster. Isolated individuals,
i.e. single-member clusters, are represented by points. Where these
circles can be placed on the page so that the distance between them
is proportional to the distance between the clusters they represent, then
they are joined by a solid straight line. In cases where the circles can
only be placed so that the distance between them is 'stretched' compared with the distance between the clusters they represent, they are
joined by a line consisting of a solid portion representing the actual

61

Table 3.5

Cannonical variate weights

	Variate 1	*Variate 2*
1	$-0{\cdot}12$	$0{\cdot}22$
2	$-0{\cdot}02$	$-0{\cdot}68$
3	$-0{\cdot}41$	$-0{\cdot}14$
4	$0{\cdot}16$	$0{\cdot}59$
5	$1{\cdot}08$	$0{\cdot}20$
6	$-0{\cdot}04$	$-0{\cdot}05$
7	$0{\cdot}29$	$-0{\cdot}22$
8	$-0{\cdot}10$	$-0{\cdot}06$

inter-cluster distance, and dashes for the remainder. Lastly 'squeezed distances', i.e. situations in which the distance between circles on the page is less than it should be for the clusters involved, are represented by V-shaped lines of the proper length.

The construction of these diagrams requires some care and thought, and some hints given by Carmichael and Sneath (1969) are extremely useful. Following these authors' suggestions, it is usually possible to preserve all or most of the small inter-cluster distances. Since these are usually the most important relations for classification, it is perhaps no great loss that the distances between distantly related items may become somewhat distorted.

Let us now examine an example of the method.

An example of a taxometric map diagram
A set of multivariate data consisting of scores on each of five variables for each of fifty individuals were subjected to the method of cluster analysis proposed by Carmichael *et al.* (1968). Five clusters were found and the information concerning these clusters necessary to produce a taxometric map is shown in Table 3.6. The taxometric map itself if shown in Figure 3.11.

This diagram shows the main relations between the clusters in a concise and easy to interpret form. The distances from each cluster to cluster 1 are indicated without distortion in the diagram. The distance most distorted by the diagram is that between clusters 3 and 5.

One way in which these diagrams could be improved would be to incorporate a graphic means for representing the number of individuals in a cluster. An easy way to do this might be to shade the circles

Table 3.6

Information needed to produce taxometric map

1. *Cluster circle radii* (i.e. one half the distance between the two most distant individuals in the cluster):

		Cluster		
1	2	3	4	5
53	30	48	50	50

2. *Centre-to-centre distances of cluster circles* (i.e. radius of cluster A + radius of cluster B + distance between nearest neighbours in A and B):

	1	2	3	4
1				
2	471			
3	484	916		
4	600	650	450	
5	600	300	1500	900

representing the clusters different amounts, increasing amounts of shading indicating larger number of individuals in a cluster.

Summary
In this chapter several techniques have been discussed which are of use when investigating multivariate data using some form of cluster analysis. A major criticism of this type of analysis has been that the majority of clustering methods will *always* produce a set of clusters, even when the data are essentially homogeneous. The methods discussed in the previous sections may prove extremely useful aids for indicating just how 'real' and distinct the clusters found by a particular technique really are. As with the other techniques discussed in this text there are advantages in using several of them on the particular set of data under investigation. In this way distortions produced by one method may be highlighted by the others.

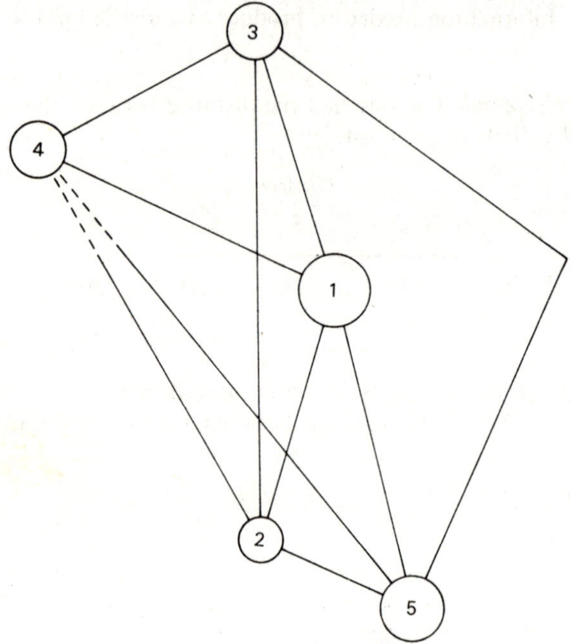

Figure 3.11   Taxometric map.

4. Some Miscellaneous Techniques

Introduction

In this chapter several further techniques for investigating multivariate data which result in some form of graphical or visual output are discussed, beginning with an account of how the well-known technique of probability plotting may be used to examine various aspects of this type of data.

Probability Plotting Techniques for Multivariate Data

The use of 'probability plots', i.e. the plotting of experimental data on probability paper, has a long and reasonably honourable history in statistics, first being suggested by Hazen (1914) in connection with a study of floods. The procedure has been used in various ways, but its main application has been to obtain a quick, informal check on distributional assumptions in the light of the sample, and to obtain rough estimates of scale and location parameters.

The basic idea behind probability plotting techniques is that a random sample from a population will tend to be representative in the sense that the empirical cumulative distribution will approximate the theoretical function. Consequently, if the sample is large enough, the sample order statistics when plotted against suitable estimates of the corresponding quantiles of the theoretical distribution will tend to yield a set of approximately collinear points about a line of slope one. The basic linearity property will remain if the sample values are subjected to a linear transformation, though the slope will no longer be one and the line will not pass through the origin.

In the following section the technique of probability plotting is described in detail and then its use in association with multivariate data indicated.

Probability plotting – technical details

Let us suppose that $y_1, y_2, \ldots, y_n$ are independent observations of a random variable Y, with a distribution function which is believed to have some particular form, say $F[(y-\mu)/\sigma]$ where μ and σ are location and scale parameters. Let $y_{(1)}, y_{(2)}, \ldots, y_{(n)}$ represent the *ordered* sample values so that $y_{(1)} \leqslant y_{(2)} \leqslant \ldots \leqslant y_{(n)}$, and now suppose that we have plausible empirical estimates, b_i, of the quantities

$$F[(y_{(i)} - \mu)/\sigma] \tag{4.1}$$

For example, a common choice for b_i is

$$b_i = i/(n+1) \tag{4.2}$$

(Other choices will be discussed later.) From the values of b_i we may, knowing the postulated form of F, obtain quantities x_i such that

$$F(x_i) = b_i \tag{4.3}$$

Now if the variable Y *does* have a distribution function of the particular form assumed, then a plot of $(x_i, y_{(i)})$, $i = 1, \ldots, n$ should be approximately linear with the intercept at $x = 0$ and the slope of the line providing rough estimates of μ and σ.

The most common type of probability plot is that involved in assessing a set of univariate data for normality. In this case the form of the distribution function assumed would be

$$F(z) = \int_{-\infty}^{z} \frac{1}{\sqrt{2\pi}} \, e^{-\frac{1}{2}t^2} dt \tag{4.4}$$

and μ and σ would be the mean and standard deviation of the distribution. For such plots the quantities x_i are easily found from tables of the standard normal distribution. To aid in the application of normal probability plotting various special types of graph paper have been produced with one scale conveniently measuring F^{-1}, so that it only becomes necessary to plot the $y_{(i)}$ against the b_i on the transformed scale.

Although normal probability plots are perhaps the most widely used, those for other distributions may also prove useful. For example, *half-normal plots* as described in Daniel (1959) and Gerson (1975) are often employed; in particular they are useful for identifying significant effects in a univariate analysis of variance (*see* Daniel (1959) for details). Such plots are also useful for investigating particular aspects of multivariate data, as we shall see on page 73. *Gamma probability plots*, as described in detail in Wilk *et al.* (1962a), are particularly useful for various investigations of multivariate data, and an example is given on page 79. This type of plot is complicated by the presence of a 'shape parameter', ξ, in the gamma distribution which must be estimated before performing a gamma plot for any reason. (Some discussion of how ξ may be estimated is given on page 78.) The computation

66

of the gamma quantiles necessary for such a plot is also fairly complicated since it involves the solution of the equation

$$\int_0^{x_i} t^{\xi-1} \exp(-t)dt = b_i \Gamma(\xi) \qquad \textbf{4.5}$$

for x_i. Full details of how to solve equation 4.5 are given in Wilk *et al.* (1962a).

The choice of plotting position, i.e. the b_i, has been discussed in some detail by Barnett (1975) and Gerson (1975). An early choice was to take

$$b_i = i/n \qquad \textbf{4.6}$$

This is, however, very unsatisfactory since problems arise with plotting the extreme observations (i.e. $i=0$ or $i=n$). Hazen (1930) suggested using

$$b_i = (i - \tfrac{1}{2})/n \qquad \textbf{4.7}$$

and this, and the value of b_i given in equation 4.2, remain in common use. Barnett (1975) indicates, however, that other choices, e.g.

$$b_i = (i - \tfrac{3}{8})/(n + \tfrac{1}{4}) \qquad \textbf{4.8}$$

or
$$b_i = (i - 0.3)/(n + 0.4) \qquad \textbf{4.9}$$

may be more satisfactory in many circumstances.

Let us now proceed to examine how the technique of probability plotting may be of use in the investigation of multivariate data.

Assessing multivariate normality
The assumption of multivariate normality underlies much of the 'classical' multivariate statistical methodology. The effects of departures from normality on the methods has only been partly investigated and the development of statistical methods that are robust to such distributional departures is still at a very early stage. Because of this it would be useful to have procedures for checking whether the assumption of multivariate normality is or is not reasonable for a given set of data. Such a test would be useful in guiding the subsequent analysis of the data, perhaps by suggesting the need for, and the nature of, a transformation to make it more nearly normally distributed, or perhaps by indicating appropriate modification of the models and methods for analysing the data. Several authors, e.g. Andrews *et al.* (1973), Healey (1968), and Mardia (1975), consider the problem and suggest various methods for checking the multivariate normality assumption. Here we

67

shall discuss a fairly simple approach based on the method of probability plotting.

This method uses the generalized distance, d_i, of each point from the sample mean vector, where

$$d_i = (\mathbf{x}_i - \bar{\mathbf{x}})' \mathbf{S}^{-1} (\mathbf{x}_i - \bar{\mathbf{x}}) \qquad \textbf{4.10}$$

and $\mathbf{x}_i$ is vector of variable values for the i-th observation, $\bar{\mathbf{x}}$ is the multivariate sample mean vector, and $\mathbf{S}$ is the sample variance–covariance matrix. If the data *are* multivariate normal then these distances have, approximately, a chi-square distribution with p degrees of freedom (where, of course, p is the number of variables). Consequently a chi-square probability plot (i.e. a gamma plot with shape parameter $\xi = p/2$) of the *ordered* distances should result in a straight line through the origin. Departures from multivariate normality will be indicated by departures from linearity in this plot.

Let us now look at some examples of this technique, beginning with its application to a set of data known to be multivariate normal.

1. *Multivariate normal data*
The first set of data investigated consisted of 200 observations sampled from a five-variate normal distribution. The resulting chi-square plot is shown in Figure 4.1. This plot is approximately linear, indicating correctly that these data are multivariate normally distributed.

2. *Multivariate data containing outliers*
A set of five variate data was generated to contain several 'wild' or outlying observations. The resulting chi-square plot is shown in Figure 4.2. The presence of a number of outlying observations is clearly indicated. These observations might now be removed and the remaining data reassessed for normality by a further chi-square plot.

3. *Non-multivariate normal data*
A set of five variate data was generated in which each variable had an exponential distribution. Two hundred such observations were obtained and the resulting chi-square plot is shown in Figure 4.3. This plot clearly deviates from linearity, indicating correctly that these data do not have a multivariate normal distribution.

4. *A set of psychiatric data*
These data consisted of scores on each of four variables for each of 180 psychiatric patients. The data were to be subjected to various types of analyses depending for their validity on the assumption of multivariate normality, and so as part of an initial screening they underwent a chi-square probability plot with the result shown in Figure 4.4.

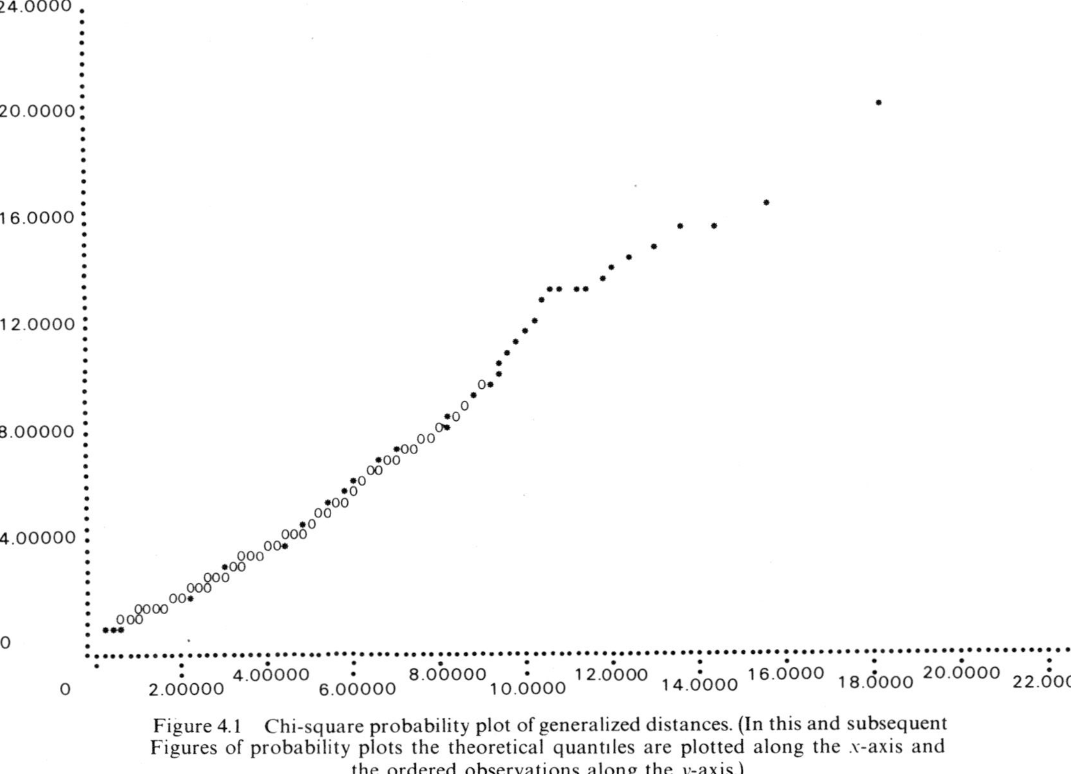

Figure 4.1 Chi-square probability plot of generalized distances. (In this and subsequent Figures of probability plots the theoretical quantiles are plotted along the x-axis and the ordered observations along the y-axis.)

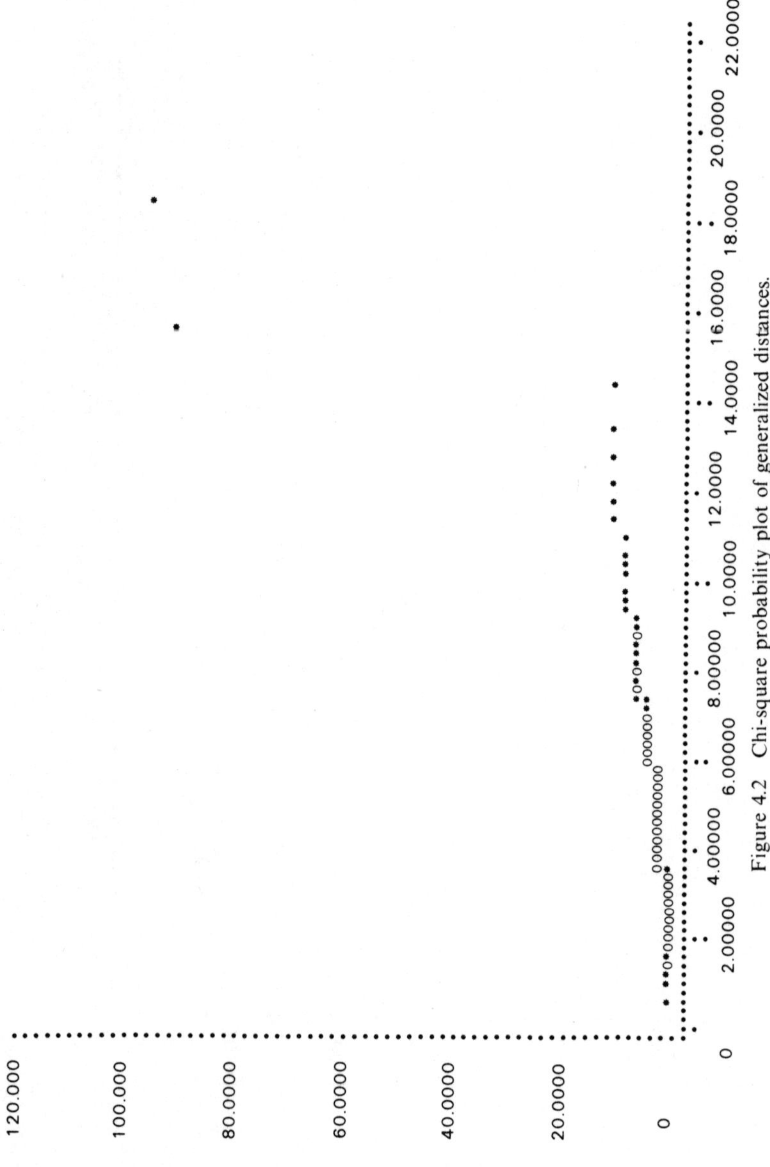

Figure 4.2   Chi-square probability plot of generalized distances.

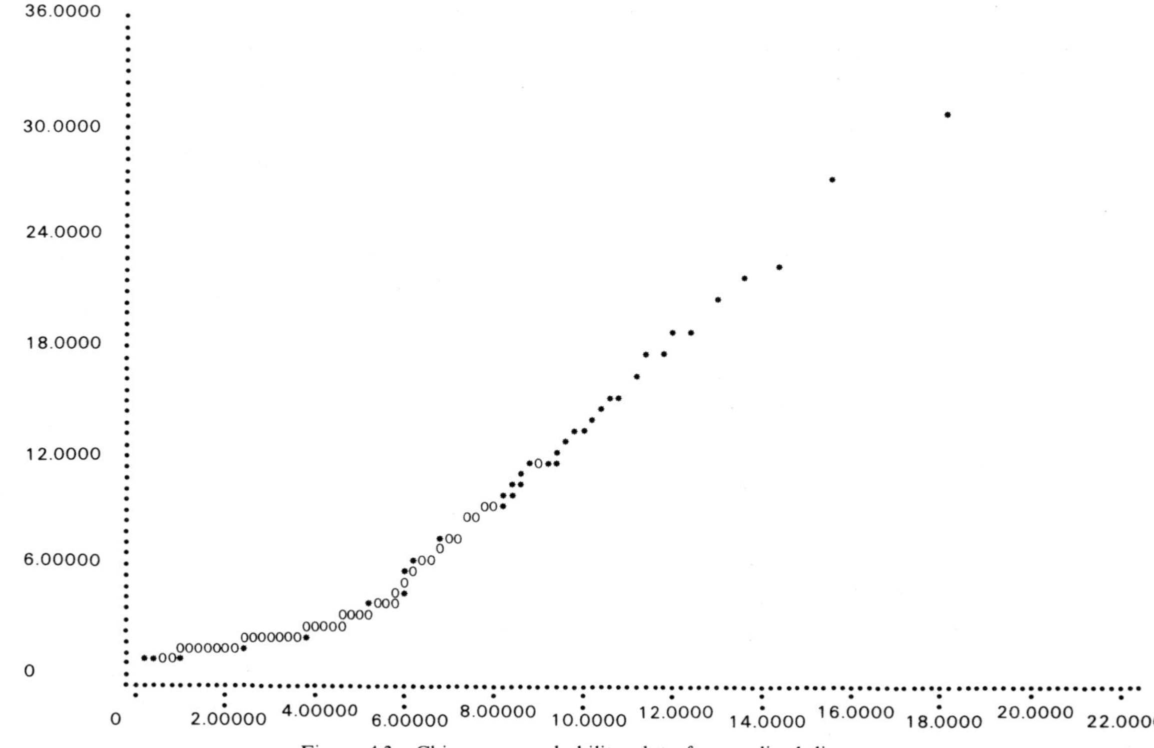

Figure 4.3   Chi-square probability plot of generalized distances.

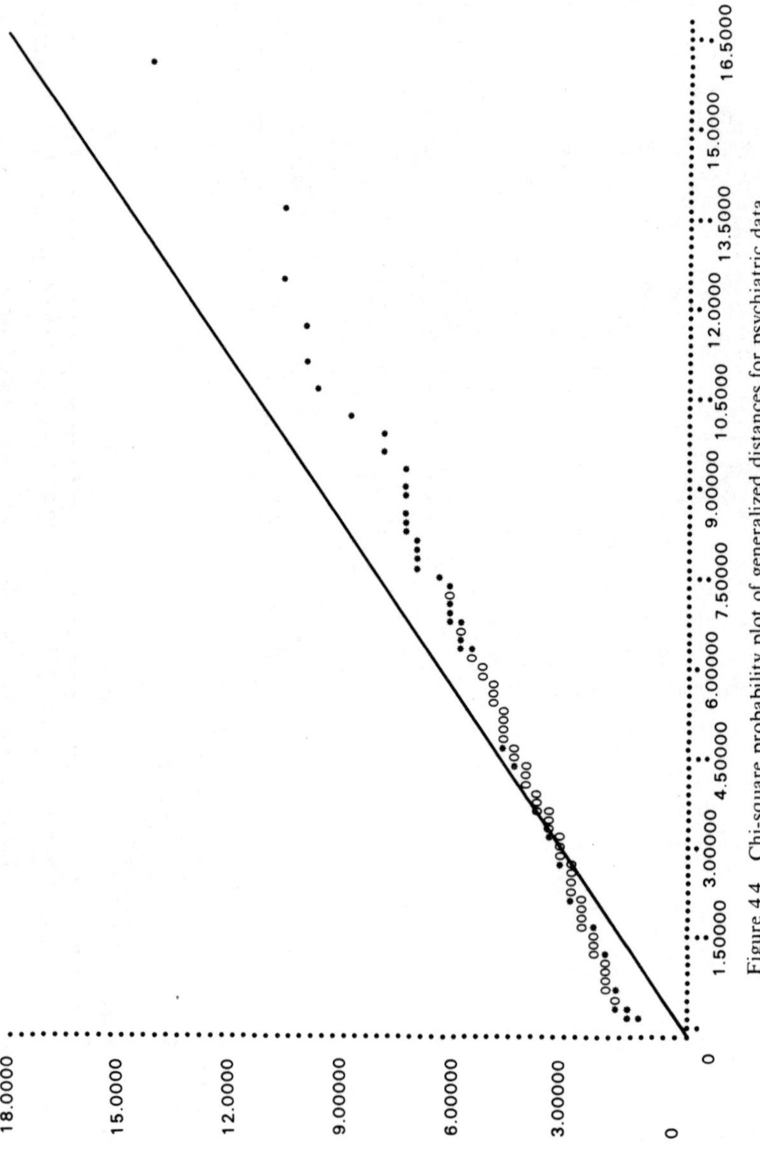

Figure 4.4   Chi-square probability plot of generalized distances for psychiatric data.

This plot shows a clear departure from linearity, indicating that these data are not multivariate normal. More detailed examination of the plot shows that there are fewer 'extreme' (i.e. large or small) distances than would be expected in the case of multivariate normal data. Such a result could be due to the presence in the data of clusters of observations, and it may be necessary to investigate this further before proceeding with other analyses.

Examining correlation matrices by means of probability plots
The starting point of several types of multivariate analysis is the matrix of correlations between all pairs of variables. Probability plotting techniques may be used on the elements of such a matrix to indicate the presence or otherwise of 'significant' (in the statistical sense only) correlations. Hills (1969) suggests performing a half-normal plot on the absolute values of Fisher's z transformation of the correlation coefficient. If the population correlation coefficient is zero then Fisher's z, given by

$$z = \tfrac{1}{2}\log_e \frac{1-r}{1+r} \qquad\qquad \textbf{4.11}$$

where r is the sample correlation coefficient, is known to have a normal distribution with mean zero and variance $1/(n-3)$. Consequently if all the $\tfrac{1}{2}p(p-1)$ correlations between the p variables are zero a half-normal plot of the absolute value of the z's corresponding to each of the coefficients should result in a straight line through the origin with slope $1/\sqrt{n-3}$.

The effect of dependence between the correlation coefficients is difficult to assess. However, since the method is graphical and in no way 'exact' it seems unlikely that its useful properties will be disturbed by the sort of dependence found in a correlation matrix.

Let us now examine some examples of this application of probability plotting.

1. *Data sampled from a population with zero correlations*
The first example arises from sampling 100 observations from a ten-variate normal distribution with zero correlation between each pair of variables. The resulting sample correlation matrix was then subjected to a half-normal probability plot, giving the result shown in Figure 4.5. This plot is clearly linear and indicates that none of the correlations are statistically significant.

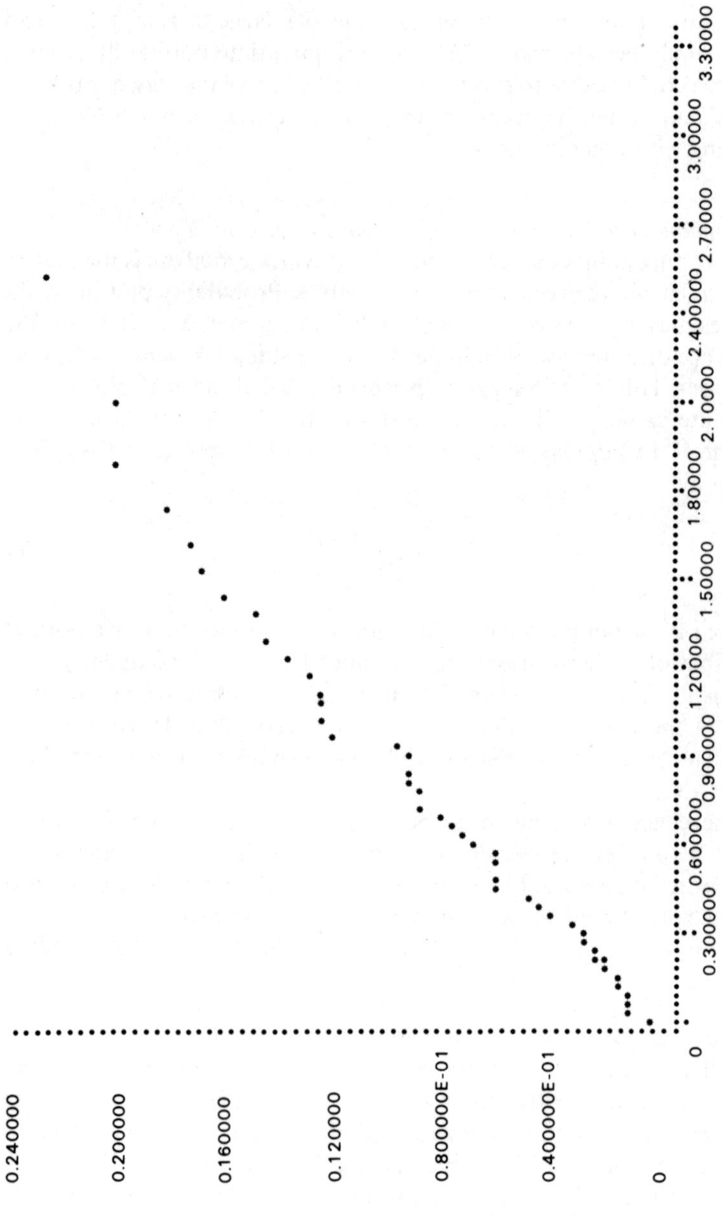

Figure 4.5   Half-normal plot of correlations.

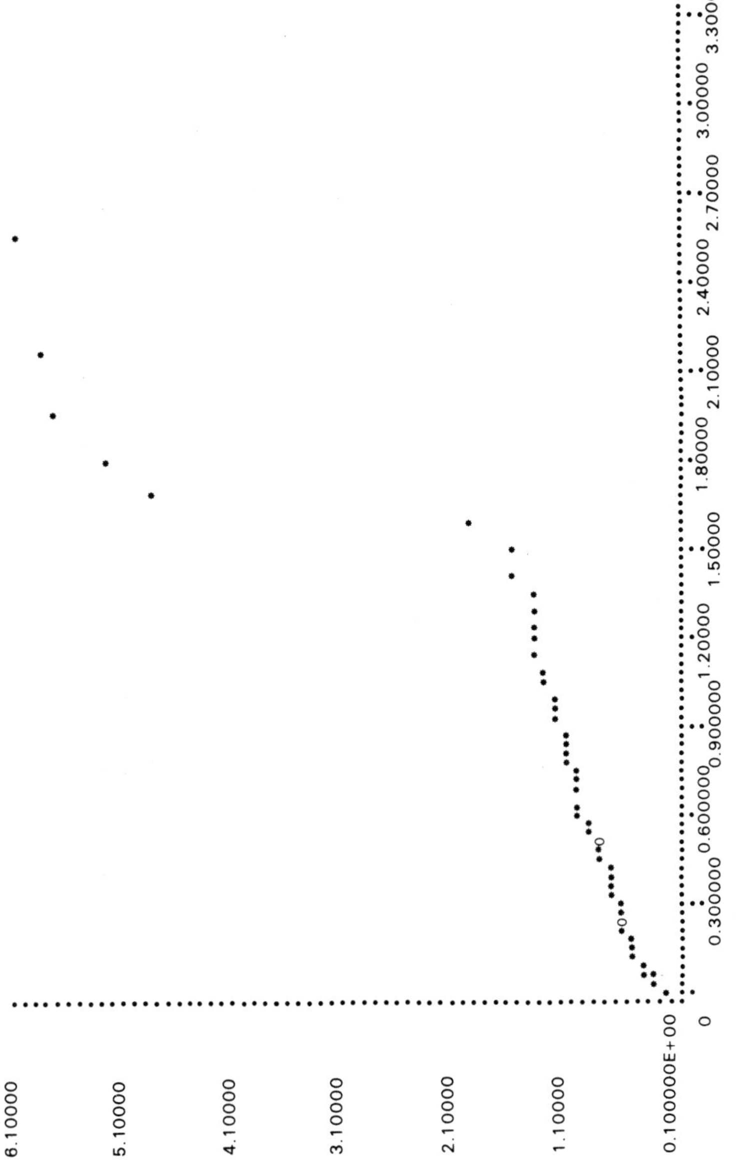

Figure 4.6   Half-normal plot of *all* correlations in example 2.

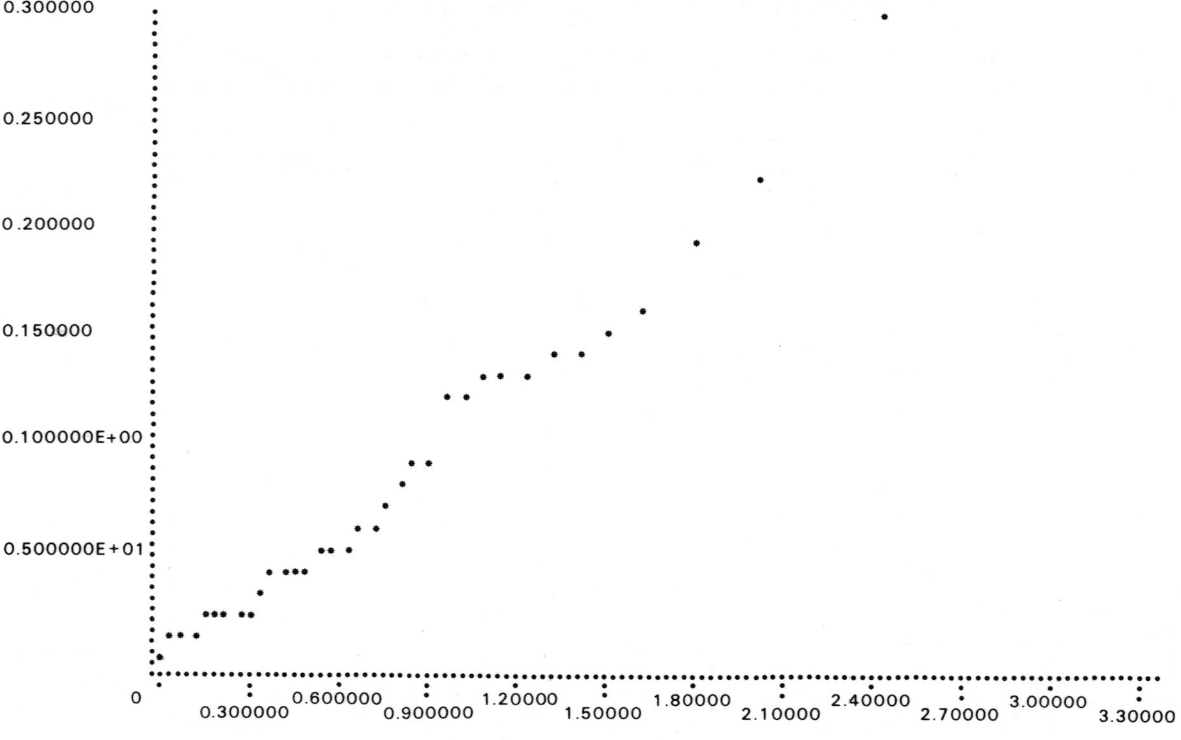

Figure 4.7   Half-normal plot of reduced number of correlations from example 2.

2. *A set of psychiatric data*

As a further illustration of the method it was applied to the correlations between ten variables measured on eighty-four psychiatric patients. The resulting plot appears in Figure 4.6. This shows clearly a departure from linearity, indicating that there are several significant correlations. If the correlations corresponding to the last ten points of Figure 4.6 are removed and a further half-normal plot is made for the remaining thirty-five coefficients we obtain Figure 4.7. This plot is approximately linear with slope 0.1, indicating that these remaining correlations are not significantly different from zero.

Probability plotting and multivariate analysis of variance

Wilk and Gnanadesikan (1964), Gnanadesikan and Lee (1970), and Gnanadesikan and Wilk (1970) describe various ways in which the technique of probability plotting may be used in the context of multivariate analysis of variance. Much of this work is too technical to be included here, and consequently we shall restrict our attention to a relatively simple application of these techniques, namely that of using gamma probability plots for identifying non-zero effects in a 2^k factorial experiment; i.e. an experiment involving k factors each at two levels.

In a multivariate situation p responses are observed on each experimental unit, and the outcome of the entire experiment may be written as an $(N \times p)$ matrix, where $N = 2^k$. When analysing an individual response in such an experiment, it is usual and meaningful to transform, orthogonally, the N observations for each response to a set of $(N-1)$ contrasts representing main effects and interactions, plus one number expressing the general mean. These $(N-1)$ contrast vectors may be visualized as $(N-1)$ points in p-dimensions, and by defining a metric in this space, a length, namely a distance from the origin, may be associated with each of the contrasts. Wilk and Gnanadesikan (1964) consider metrics of the form.

$$d = \mathbf{x}'\mathbf{A}\mathbf{x} \qquad\qquad \textbf{4.12}$$

where $\mathbf{x}$ represents one of the p-dimensional contrast vectors and $\mathbf{A}$ is some chosen *compounding* matrix which converts the multidimensional vector $\mathbf{x}$ into a single number. Some examples of appropriate compounding matrices are

1. A diagonal matrix of reciprocals of variances (specified or estimated) of the p responses.

2. The inverse of the covariance matrix (specified or estimated) of the responses (cf. Mahalanobis distance).

More is said concerning the choice of compounding matrix later.

For fixed $\mathbf{A}$, Gnanadesikan and Wilk show that under the assumption of no real treatment effects, the $(N-1)$ distances $d_1, d_2, \ldots, d_{N-1}$ corresponding to the $(N-1)$ contrast vectors may, as a reasonable approximation, be expected to behave like a random sample from a gamma distribution with origin parameter zero, unknown scale parameter λ, and unknown shape parameter ξ. If a suitable estimate of ξ were available then a meaningful summary of the experiment could be obtained by plotting the $(N-1)$ ordered distances against the appropriate quantiles of the standard (i.e. $\lambda=1$) gamma distribution using the procedure suggested by Wilk *et al.* (1962a). If the null assumption is correct, and ξ properly chosen, then the resulting plot should tend to be a straight line passing through the origin with slope $1/\lambda$. Non-zero effects will be indicated by a departure from linearity. (This procedure depends *only* on having an estimate of ξ; the scaling factor λ influences the slope and *not* the collinearity of the points.)

A suitable procedure for estimating ξ is described in Wilk *et al.* (1962b). This estimation procedure operates on a number, say the first K, of the ordered distances, where $K<(N-1)$. It is usually reasonable to assume that these smallest distances do essentially satisfy null assumptions. The inclusion of *all* the distances in the process of estimating ξ would tend to generate a value for ξ which would obscure the fact that some of the distances are statistically 'too large', since the larger distances may not obey null assumptions. Wilk *et al.* (1962b) indicate that the estimation process is rather insensitive to the choice of K provided this is not too close to $(N-1)$.

A question which arises with this technique is in respect of the choice of compounding matrix, $\mathbf{A}$. Wilk and Gnanadesikan recommend that, in general, several different compounding matrices should be employed They reason that the majority of multidimensional situations cannot be entirely described in a single one-dimensional representation, and that each distinct distance function plot (corresponding to the different choices of $\mathbf{A}$), may give a different insight into the data.

As with the other applications of probability plotting described in previous sections, replotting may in some cases be desirable; e.g. if an initial plot indicates evidence of some real effects then their existence will affect the entire configuration; replotting after omission of the

Table 4.1

Contrast vectors and compounding matrix for gamma probability plot example

		Variable 1	Variable 2
	A	−1·387	−0·787
	B	−2·037	−0·704
	C	0·096	0·712
Contrast	D	−0·820	−0·371
	AB	9·458	7·758
	AC	1·191	−0·408
	AD	−1·708	−1·342
	BC	1·392	1·292
	BD	0·825	−0.275
	CD	0·125	0·525
	ABC	−1·550	0·050
	ABD	−2·950	−1·950
	ACD	0·450	0·450
	BCD	0·117	0·050
	ABCD	−1·633	0·900

$$A = \begin{bmatrix} 0·204 & 0·124 \\ 0·124 & 0·265 \end{bmatrix}$$

points corresponding to these may enable other more subtle effects to be observed.

Let us now look at a simple example of this technique.

A gamma probability plot for a 2^4 factorial experiment
In this experiment four factors (A, B, C, and D) were involved. In each cell of the experiment there were three bivariate observations. The estimated contrasts for main effects and interactions for each of the two response variables are shown in Table 4.1. The compounding matrix, A, chosen for these data was the pooled within-cell variance–covariance matrix (see Table 4.1). Using the first ten of the ordered distances the shape parameter, ξ, was estimated to be 0·74. The gamma probability plot of the fifteen distances using this value of ξ is shown in Figure 4.8. This figure clearly indicates that there is a large interaction effect between factors A and B. The assessment of the remaining effects is not possible on this plot, and in any detailed investigation of these data a

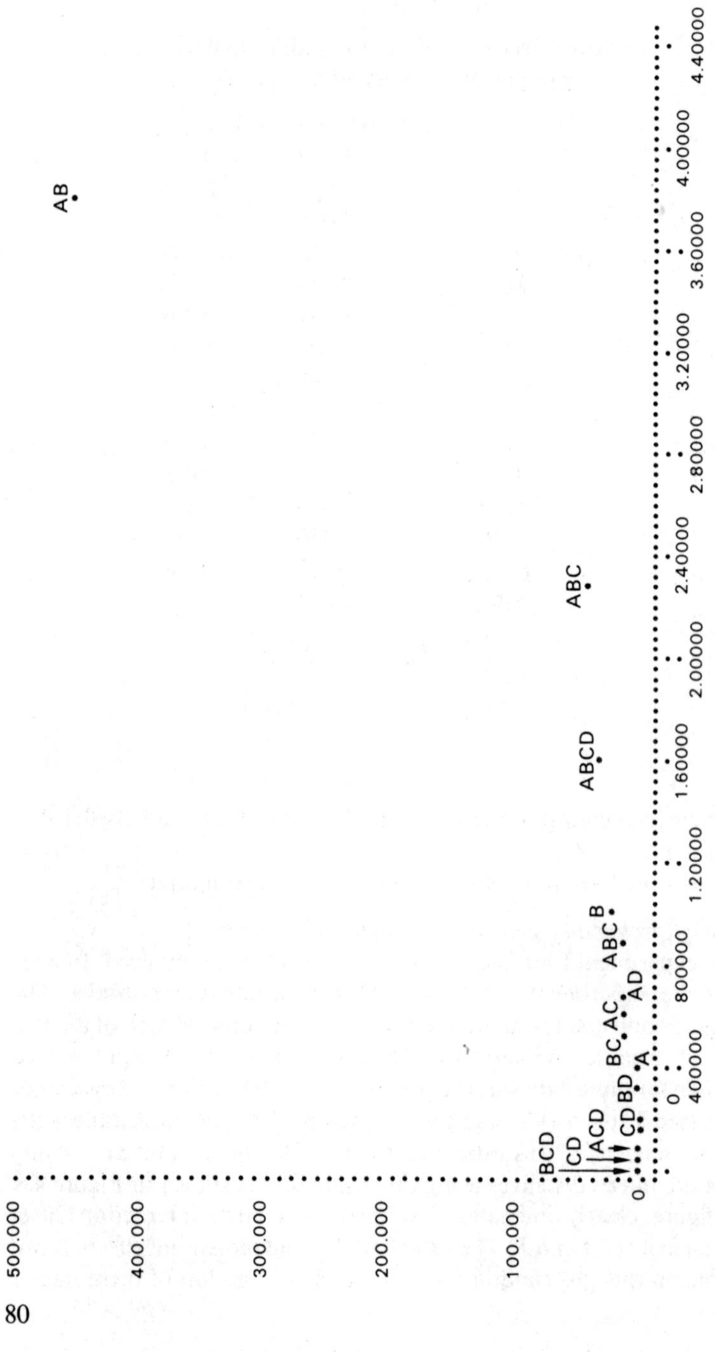

Figure 4.8 Gamma probability plot of bivariate contrast vectors.

further plot would be necessary of the fourteen distances that remain after that corresponding to the *AB* interaction has been removed.

Some further interesting possibilities for the use of probability techniques with multivariate data are described in Gnanadesikan and Kettenring (1972). In particular these authors indicate how such techniques might be valuable in the analysis of multidimensional residuals.

Andrews' Plots of Multivariate Data

Andrews (1972) proposes a very simple technique for obtaining a visual representation of multivariate data in which each multivariate point is mapped into a function of a particular form. This technique has been found useful for identifying clusters, outliers, etc. in multivariate data as we shall see in the examples given later. We begin, however, with a detailed description of the method.

Andrews' plots – technical details

Andrews' procedure is, essentially, very simple; each of the p-dimensional observations, **x**, defines a function.

$$f_{\mathbf{x}}(t) = x_1/\sqrt{2} + x_2 \sin(t) + x_3 \cos(t) + x_4 \sin(2t) + x_5 \cos(2t) + \ldots \qquad \textbf{4.13}$$

This function is then plotted over the range $-\pi \leqslant t \leqslant \pi$. A set of p-dimensional observations will now appear as a set of lines drawn across the plot. Andrews shows that this particular function has many properties which make it particularly useful in the exploration of multivariate data. Perhaps the most importance of these is that this function representation preserves Euclidean distances. Consequently points which lie close together in the original p-dimensional space of the observations will be represented by lines on the plot that remain close together for all values of t, whilst distant points will be represented by lines which remain apart for at least some value of t. This property enables the plots to be used for the possible identification of cluster, outliers, or other peculiarities of the data.

Examining the form of the function in 4.13, we see that the original variables (i.e. $x_1, x_2, \ldots, x_p$) are not equally weighted because some of them are associated with cyclic components which have a high frequency, others with components having a low frequency. Since in these plots low frequencies are more readily seen than high frequencies, it may be useful to associate x_1 with the variable considered most important, x_2 with the second most important, and so on. In the absence

81

of any firm ideas as to such an order of variables it may be useful to perform a principal components analysis prior to the Andrews' plotting, now associating x_1 with the first component, and so on.

It can also be useful to examine the function values associated with a particular value of t, say t_0. Andrews shows that $f_x(t_0)$ is proportional to the length of the projection of the vector **x** on to the vector

$$f_1(t_0) = (1/\sqrt{2}, \sin(t_0), \cos(t_0), \sin(2t_0), \cos(2t_0), \ldots) \qquad \textbf{4.14}$$

The projection onto this one-dimensional space may reveal patterns that occur only in this subspace and which may be otherwise obscured by other dimensions.

A problem which arises when using this technique is that only a fairly limited number of observations may be plotted on the same diagram before this becomes too confused to be helpful. When many observations are to be plotted, the most useful procedure is to first plot all of them on one graph to assess the general characteristics of the sample, and then make separate plots of, say, sets of ten observations; these may be used to consider individual observations in relation to the whole.

Andrews considers questions of significance and confidence limits for these plots assuming that the p-variables are uncorrelated. Goodchild and Vijayan (1974) extend the significance tests arrived at by Andrews to the situation where such an assumption is not justified. These tests will not, however, be considered here since they are against the spirit of the informal data exploration approach of this text.

Let us now examine some examples of the application of Andrews' plotting procedure to multivariate data.

Using Andrews' plots to identify clusters
Figure 4.9 shows Andrews' plots of thirty, five-dimensional observations. Examining the diagram at $t = 0.5$, we see that there are three well-separated 'bands' each containing the plots of ten observations. This clearly indicates that these data consist of three well-separated clusters of observations.

Andrews' plots of some data on Indian castes and tribes
These data are taken from Mahalanobis *et al.* (1949), Table 8.11, and consist of eight cannonical variate means for each of twenty-two Indian castes and tribes. These eight means were derived from measurements of twelve quantitative variables such as head length, head breadth, etc. on samples of about 160 individuals from each of the

82

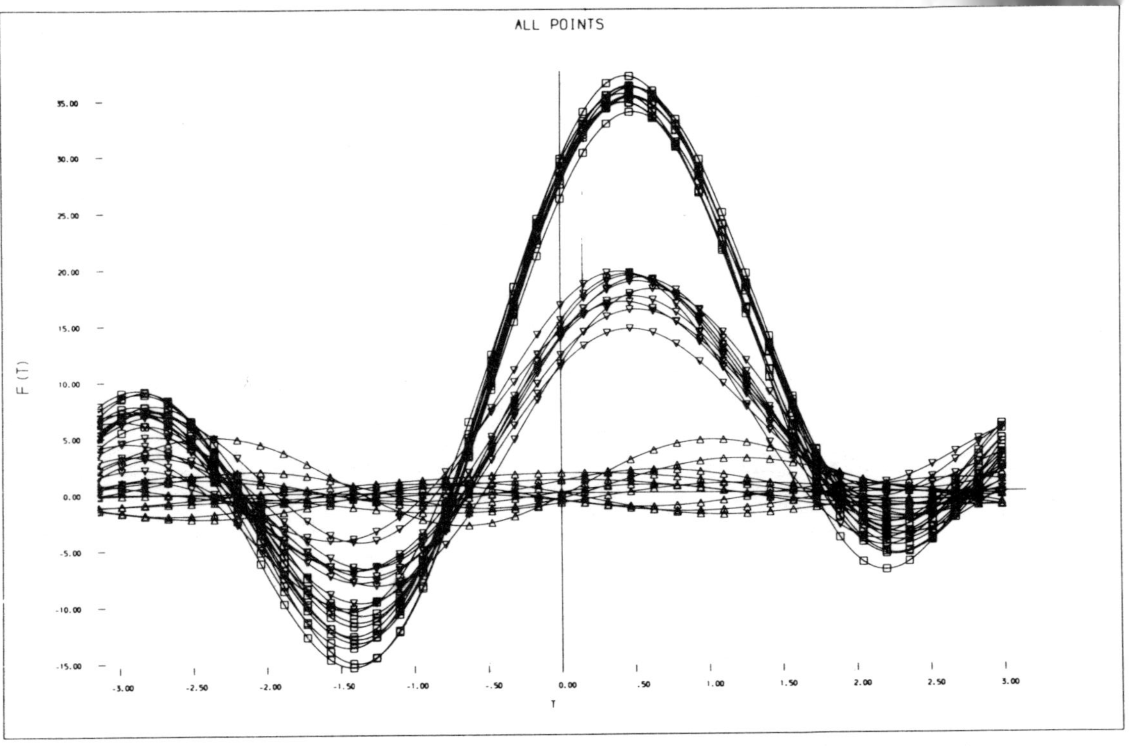

Figure 4.9   Andrews' plots for a set of thirty, five-dimensional observations generated to contain three clusters.

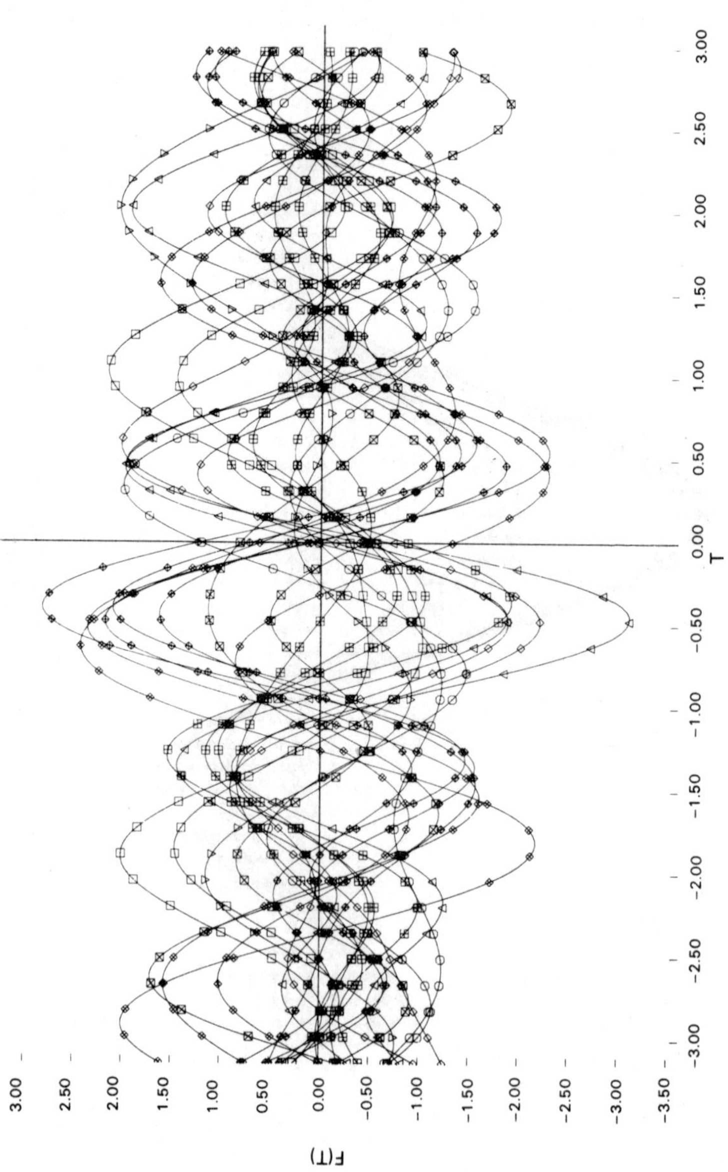

Figure 4.10 Andrews' plots of all twenty-two Indian tribes.

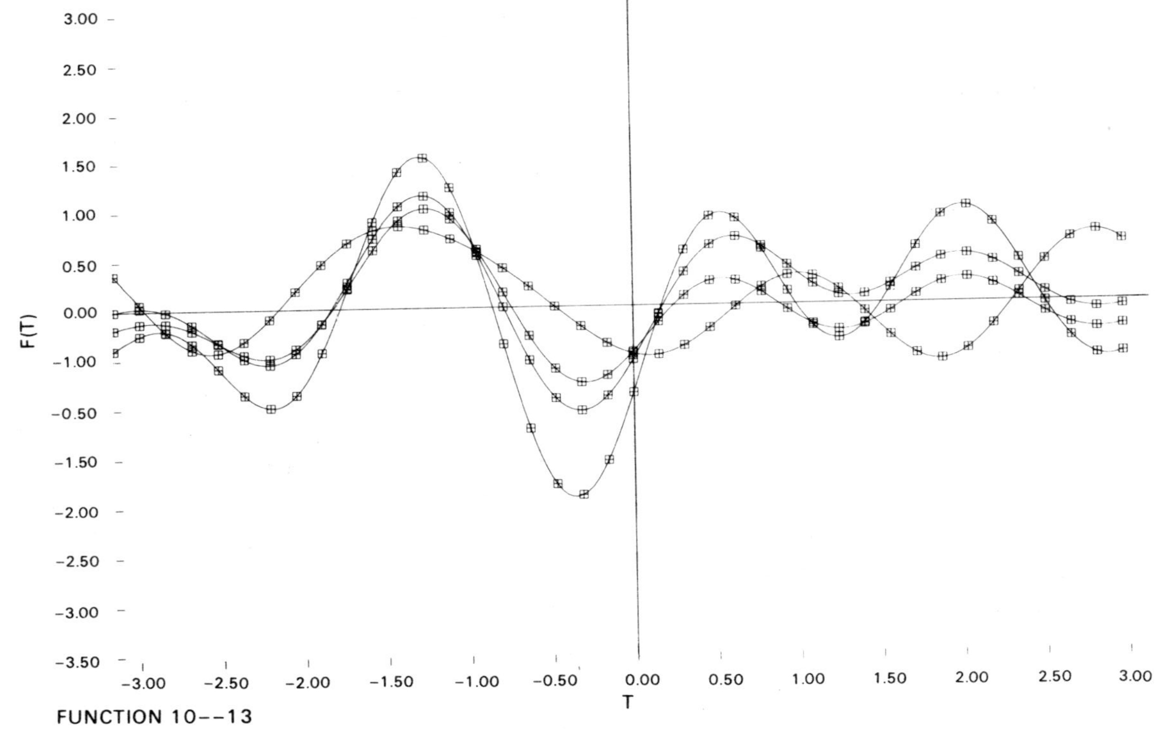

Figure 4.11   Andrews' plots of Artisan cluster of Indian tribes.

85

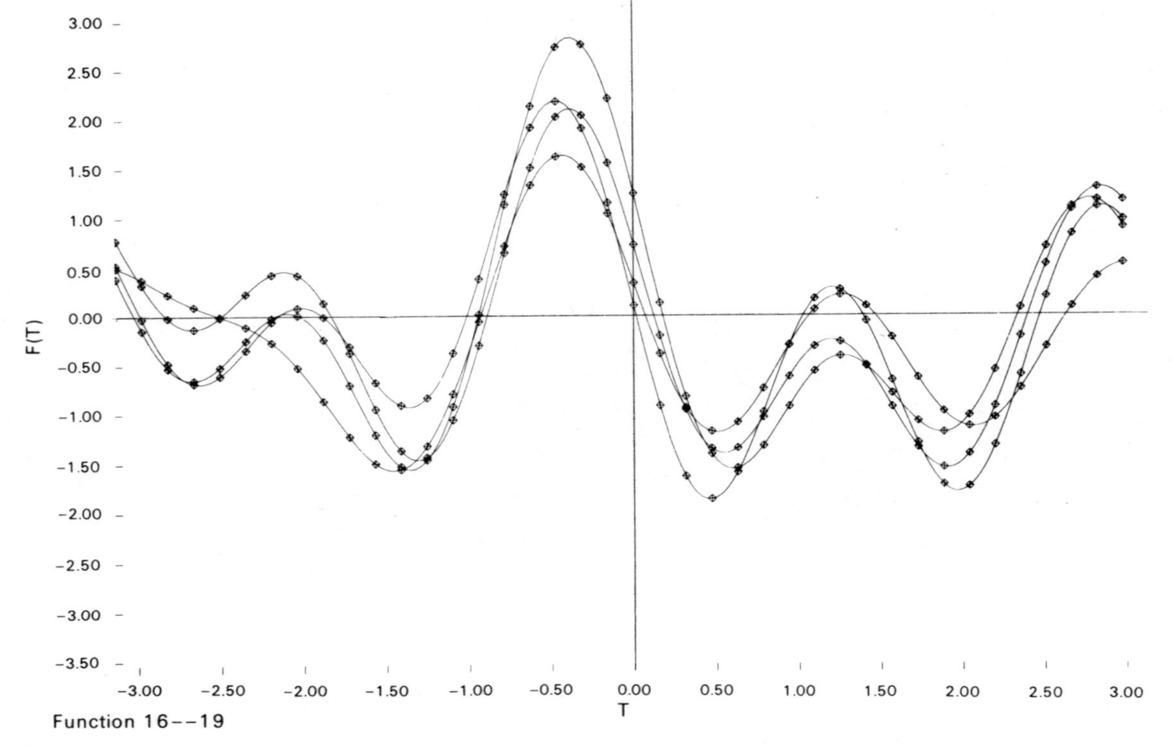

Figure 4.12   Andrews' plots of Chero, Majhi, Panika, and Kharwar tribes.

twenty-two tribal populations. (These data have been considered previously in Chapter 3.) Figure 4.10 shows Andrews' plots for all twenty-two observations.

This diagram, whilst obviously of no use for detailed study, does indicate that there are large differences between the tribes and that in no way may these be considered a homogeneous set of data. To obtain more information about particular tribes or subsets of tribes we need to make separate graphs containing only those plots of interest. For example, Figure 4.11 shows the plots of four of the tribes, namely the Artisan cluster mentioned in the original work. We see that although the four plots are basically very similar, there are differences, notably around $t = -0.5$, which may be worth investigating further using the one-dimensional projections for this value of t. A further set of tribes, Chero, Majhi, Panika, and Kharwar, considered to be a cluster by Jardine and Sibson (1971), who also analysed these data, is plotted in Figure 4.12. These appear rather more homogeneous than the previous cluster, the plots remaining very close to each other for all values of t.

By plotting various subsets of the data in this way, it is generally possible to build up a fairly detailed view of the data under investigation.

Representing Multivariate Data as Faces

Chernoff (1973) describes a method for representing multivariate data graphically in which each multidimensional observation is represented by the cartoon of a face, the features of which are governed by the values taken by particular variables. A sample of multivariate observations is now represented by a collection of such faces, and these may then be examined in a bid to identify clusters, outliers, etc. In the following section the technique is described in more detail.

Faces – technical details
Frith (1973) has produced a computer program for drawing faces to represent multivariate observations and it is his method which will be described here and used in the examples that follow. This method consists of representing the multivariate observation by a face which lies somewhere between two 'extreme' faces, these being shown in Figures 4.13 and 4.14. The particular face defined by an observation depends on that observation's variable values and is determined by interpolating

between the two extreme faces on each of nine features, each feature being associated with one of the variables. The features which the variables determine are shown in Table 4.2. More than nine variables could be accommodated simply by adding other features.

Of course, the observations could just as easily be represented not as faces but as other 'objects'. For example, Figure 4.15 compares the

Table 4.2

Cartoon face features

1. Upper hair
2. Chin curve
3. Lower hair
4. Eye size
5. Mouth size
6. Eye space
7. Eye slant
8. Mouth curve
9. Face size
(eyes to mouth)

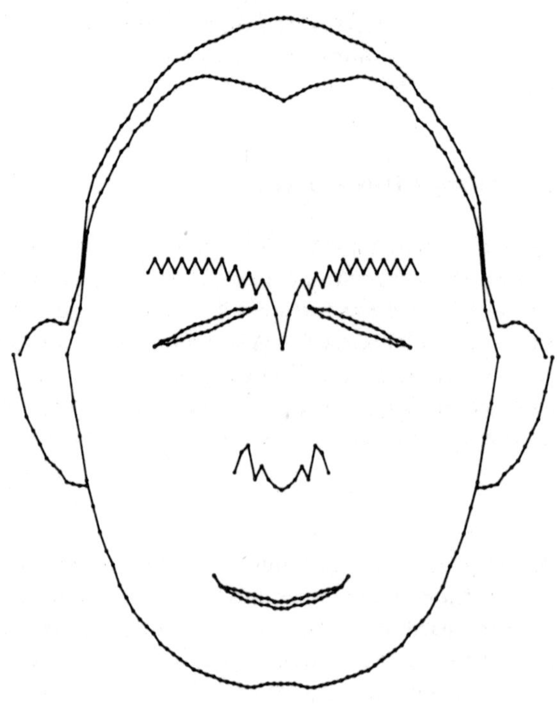

Figure 4.13 'Extreme' cartoon face.

faces representations with two other possibilities. However, the representation in terms of faces may be more useful than others since people are used to studying and reacting to faces and, hopefully, they are able to filter out insignificant visual phenomena and focus on the potentially important. Certain major characteristics of the faces are instantly observed, and finer details become apparent after studying the faces for some time. The awareness of these does not drive out of mind the original major impressions.

Certain questions arise with respect to this method for representing multivariate data graphically. The first is whether some features of the face are more informative than others? The second is whether different observers use different features of the faces to judge their similarity, etc.? The answer to both these questions would seem intuitively to be yes. Certain observers may concentrate on, say, the eyes, others on the chin, etc. Certain features are more readily seen than others (see example below) and so these will obviously be more informative. Both situations

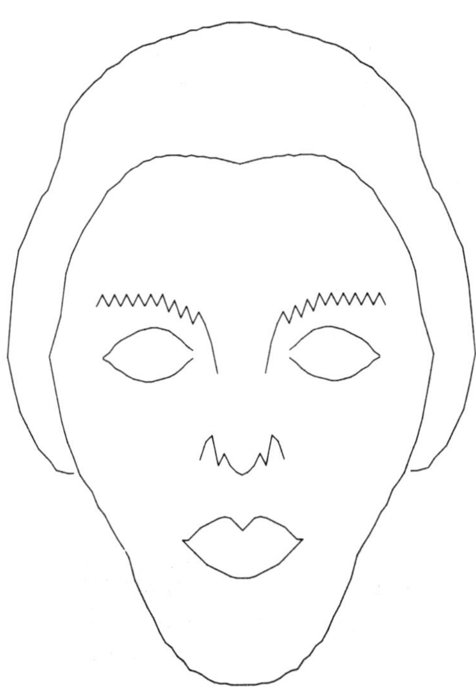

Figure 4.14 'Extreme' cartoon face.

raise the issue of how subjective will be an observer's judgement about the faces? It would appear likely that a large amount of subjectivity may be present, although in Chernoff's opinion humans will tend to concentrate on features of the face that reflect what is important in the data. A possible procedure for attempting to overcome such problems is to produce several sets of faces for the data, each set involving a different permutation of variables to features. An example of how such a procedure may be helpful is given on page 93.

Let us now examine some applications of the faces technique.

Faces representation of some data containing clusters
The first example of this technique is on a set of nine variate data generated to contain three well-separated clusters. Thirty observations were taken and the resulting thirty faces are shown in Figure 4.16. These faces clearly indicate the presence of the clusters in the original data.

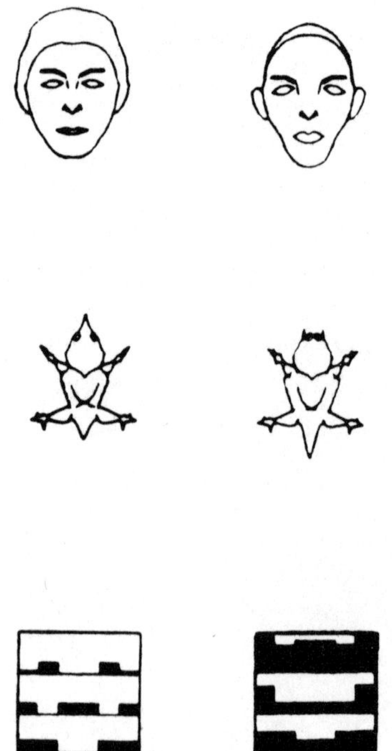

Figure 4.15   Faces and other possible alternative representations of multivariate data.

Faces representation of some data on ships
The data for this investigation were extracted from *Jane's Fighting Ships*, and consisted of various variables recorded for each of a number of ships. The variables used were as shown in Table 4.3. The twenty-five ships used in the investigation consisted of a number of ships from various classes, as shown in Table 4.4.

The major interest in using the faces technique on these data was to see how well it would indicate the structure present, i.e. identify that there were different types of ships. (Before applying the technique to these data they were standardized to zero mean and unit variance.) Figure 4.17 shows the faces representation of these data when the correspondence between variables and features is as indicated in Table 4.3.

Figure 4.16   Faces representation of thirty, nine-dimensional observations generated to contain three clusters.

Table 4.3

Correspondence between ship variables and face features for Figure 4.17 (the numbers refer to the features as given in Table 4.2)

1. Displacement in tons
2. Length in feet
3. Beam in feet
4. Number of heavy guns
5. Number of light guns
6. Ship's complement
7. Maximum speed
8. Submersibility
9. Number of aircraft carried

Table 4.4

Ship types

Ships 1–5: Aircraft carriers   *Ships 6–9:* Destroyers
Ships 10–19: Submarines   *Ships 20–25:* Frigates

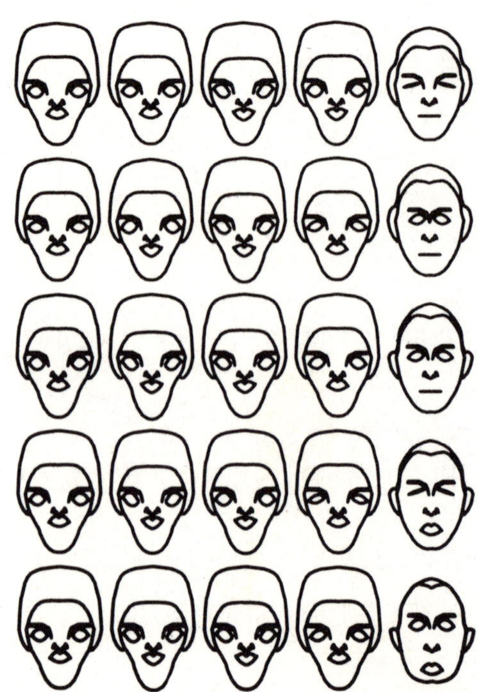

Figure 4.17 Faces representation of ships' data when correspondence between features and variables is as indicated by Table 4.3.

(To identify which face represents which ship, Figure 4.17 should be read from bottom to top along columns, starting with the column on the right-hand side, so that the face in the bottom right-hand corner represents ship number 1, etc.) The first impression gained from this set of faces is that only those representing aircraft carriers are clearly distinguishable.

Of these five faces, two are considerably different from the others. Since two of the aircraft carriers in this study were large American ones, and the others were rather smaller British ships, this would appear to be reasonable. On closer examination of the remaining twenty faces, the most obvious distinguishable feature is mouth curve, which is here representing the variable 'submersibility'. Those faces representing submarines are 'smiling', those representing frigates and destroyers are not! The faces representing frigates and destroyers are virtually indistinguishable.

A further collection of faces was constructed for this set of data using the association between variables and features indicated by Table 4.5. These faces are shown in Figure 4.18. The structure of the data is now more clearly seen. Aircraft carriers and submarines are clearly distinguishable. At first glance the faces representing frigates and destroyers appear similar but closer inspection shows that the size of the mouth is different, being larger for those faces representing frigates. Mouth size is, in this case, representing the variable 'maximum speed' and frigates and destroyers have similar properties except that destroyers are generally faster. In Figure 4.17 this variable was represented by the feature 'eye slant' which, since the eyes were large for the faces representing both frigates and destroyers, was not very helpful in distinguishing between the two classes.

Some other applications of this method are reported by Chernoff who, in addition to using the technique to identify clusters, etc., also uses faces for detecting the time points at which a multivariate stochastic process changes character by looking at a sequence of faces corresponding to successive points in time to locate the places where the faces change their character.

Table 4.5

Correspondence between ship variables and
face features for Figure 4.18

1. Number of aircraft	2. Submersibility	3. Number of light guns
4. Beam	5. Maximum speed	6. Length
7. Number of heavy guns	8. Ship's complement	9. Displacement

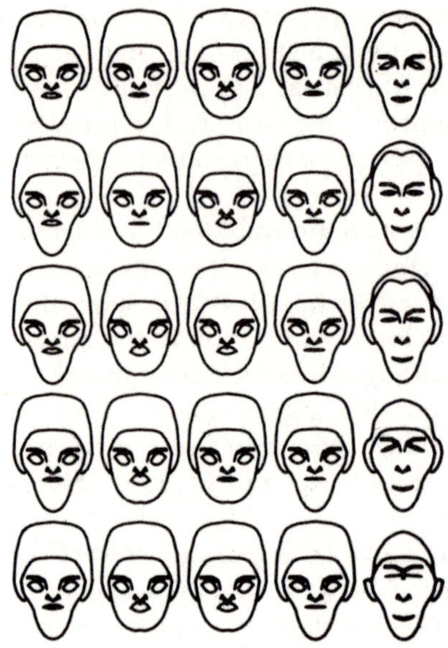

Figure 4.18 Faces representation of ships' data when correspondence between features and variables is as indicated by Table 4.5.

Summary

In this chapter various techniques which may be used for a graphical representation of multivariate data have been described. All have the considerable advantage of requiring very little computational effort. There are, of course, many other techniques which might have been included here. For example, each observation could be represented by a type of 'bar profile' of the type shown in Figure 4.15, or by a similar diagram in which the bars are replaced by a polygonal line. Such profile diagrams are in fairly common use but probably do not possess the 'penetrating' power of the methods discussed in this chapter. Chernoff (1973) describes a much more effective variation of the profile method in which polygons are used to represent the observations. Again, however, the methods discussed in the previous sections would appear more helpful and informative.

Finally it needs to be stressed once again that no claims are made that the techniques described in this chapter (or those described in previous chapters) provide more than a partial solution to dealing with the complexities of multivariate data. Nevertheless weighed against the computational effort they require they do, in many cases, provide useful and informative insights into the structure of such data.

5. Some Final Comments

Introduction

In the previous chapters many techniques which allow a visual representation of multivariate data have been described. The main purpose of such techniques is to afford an informal exploration of this type of data. In this chapter we shall briefly reiterate some of the points made previously with regard to this informal approach to dealing with multivariate data, and also discuss some of the problems such as dealing with 'large' data sets.

Informal and Formal Inference

The techniques discussed in the body of this text are useful, essentially, for what Gnanadesikan (1973) terms 'informal inference'. By this he means procedures which allow investigators to develop a 'feel' for their data and to gain insights into the structure and the unanticipated peculiarities of that data. These techniques are to be contrasted with the more familiar processes of formal statistical inference which are concerned with precise (generally probabilistic) statements under tightly specified models. However, as already indicated in Chapter 1, the two approaches should be regarded as complementary rather than as exclusive alternatives. Gnanadesikan makes this point clearly in the following statement:

> 'For truly multiresponse problems careful, complex and extensive analyses via numerical computations and summaries would be necessary, and simple graphical representations should not be considered substitutes for such analyses but in conjunction with, and as aids for, them. Graphical tools and technology are no panacea, as indeed nothing seems to be in the real world of data analysis.'

(Such a comment should be noted carefully, particularly by those researchers who believe that the many widely available and well-documented packages for formal multivariate analyses constitute such a panacea.)

What appears to be desirable in the exploration and analysis of multivariate data is far more interaction between the two approaches of 'informal' and 'formal' inference. In part this will be made more likely once greater experience has been gained with the use of the less familiar of the graphical techniques described in this text. The fairly small computational effort needed to implement many of these methods should make this task reasonably easy.

Large Data Sets

At first sight, a grave disadvantage of many of the techniques described in the preceeding chapters is that they appear only able to deal with relatively small data sets. (In general, we mean by this small values of n, the sample size.) Many readers, having noted that several of the examples used to illustrate various of the techniques involve only twenty or so 'individuals', may even have decided that these methods have little or no relevance for their particular problems which may involve a value of n of, say, 200–500. Such a view would not necessarily be correct. Several of the examples when examined more closely are seen to involve data sets which, originally at least, would be designated by most people as 'large'. For example, the astronomical data considered in Chapter 2 consisted originally of seven physical characteristics for each of 273 galaxies. Again the data on Indian tribes and castes analysed in Chapters 3 and 4 derive from measurements of twelve quantitative attributes on about 160 individuals from each of twenty-two caste and tribal populations. The graphical techniques used on these data sets were applied *not* to the raw data but to a reduced data set resulting from some preliminary analysis. For example, the galaxies involved in the astronomical data were first grouped according to the traditional morphological system of classifying galaxies, and then an inter-group distance matrix was computed. It was this distance matrix which was then subjected to a particular graphical analysis. Using these techniques on such reduced data sets in this way can still be a very useful procedure; consequently many of the methods not able to deal *directly* with large data sets may still be used to obtain informative graphical displays after the n individuals in the sample have been subjected to a preliminary grouping of some kind (using perhaps *a priori* information or some form of cluster analysis), and a between groups distance matrix or a set of group means computed.

Of course, some of the techniques may be used directly with rather larger sets of data than those used in the examples (which have also had to be kept small for reasons of space, etc.). For example, it is possible to use principal component, principal co-ordinate, and non-linear mapping plots directly on sets of data involving two or three hundred individuals. Again there is no difficulty in using those techniques based on probability plotting on large data sets.

Since the numbers to input to the particular graphical technique involved are at the control and the discretion of the investigator, it should almost always be possible to obtain at least *some* visual display of the data, whether in its original or reduced form, at some stage of the investigation; such a display will generally be informative and of value for reasons mentioned previously, such as indicating further analyses which may be helpful, or in forming a convenient summary of some aspect of the data useful for communication, etc. Consequently the visual representation is almost certainly worth obtaining. (Perhaps at this point readers should be reminded of those two ancient maxims that 'every picture tells a story' and 'one picture is worth a thousand words'.)

Summary

The techniques described in this text for the examination of multivariate data have in common that they produce some kind of visual or graphical output. The emphasis has been on methods which do *not* as a *necessity* require *specialized* graphical hardware, although in many cases it is convenient to obtain the graphical output via some type of computer device such as a line printer or a microfilm plotter. In this text no details are given of the fast developing fields of sophisticated computer graphics and interactive graphics terminals, although these may be of great importance in the development and teaching of multivariate analysis in particular, and of course statistics in general. The reason for this is that the text has the rather limited aim of providing an introduction for research workers in many fields to some graphical techniques for dealing with multivariate data with which they may not be familiar; for the majority of such workers the more 'exotic' type of computer facility mentioned is just not available.

No claim is made that this text provides a comprehensive survey of all those techniques for handling multivariate data which might be termed graphical, although it is hoped that the majority of the most useful methods are included.

This text began with a quote from the work of one statistical 'giant', Sir R. D. Fisher, and it seems appropriate to end with a quote from another, Professor E. S. Pearson, who, whilst not *always* in agreement with Fisher, appears to be so in respect of statistical diagrams. The following is taken from Pearson's 1956 Presidential address to the Royal Statistical Society:

> 'I have little doubt that, where this is possible, a visual survey of the "pattern" of his data provides the statistician who is trained to extract a meaning from diagrams with the quickest method of checking whether the model he proposes to use is likely to be appropriate or not. That understanding can be achieved through visual aids may be regarded as a proposition so obvious that it needs no restatement by me;...'

Appendix A: The Measurement of Similarity and Distance

Many excellent accounts of the measurement of similarity and distance are available, e.g. Andeberg (1973), Sneath and Sokal (1973), and Gregson (1974), and consequently in this Appendix only a fairly brief account of the topics will be given. We begin with the measurement of similarity.

Similarity Measures

Definition: a non-negative real valued function, $s(x, y)$, of pairs of points of a set E is said to be a similarity measure for E if

1. $0 \leqslant s(x, y) \leqslant 1 \qquad x \neq y$
2. $\quad s(x, x) = 1$
3. $\quad s(x, y) = s(y, x)$

(Note that this particular definition excludes measures such as the correlation coefficient which may take negative values.)

Similarity measures are used most commonly when the data consist of variables of the 'presence' and 'absence' type. For such data many different similarity measures have been proposed, and a number of these are given in Table A.1. The terms a, b, c, and d used in this table

Table A.1

Similarity measures for presence–absence data

1. $\dfrac{a+d}{a+b+c+d}$ 2. $\dfrac{a}{a+b+c+d}$

3. $\dfrac{a}{a+b+c}$ 4. $\dfrac{2a}{2a+b+c}$

5. $\dfrac{2(a+d)}{2(a+d)+b+c}$ 6. $\dfrac{a+d}{a+d+2(b+c)}$

7. $\dfrac{a}{a+d+2(b+c)}$ 8. $\dfrac{a}{a+2(b+c)}$

9. $\dfrac{a}{b+c}$ 10. $\dfrac{a+d}{b+c}$

99

refer to the entries in the familiar two-way association table for a pair of individuals, namely

		Individual i		
		$+$	$-$	
	$+$	a	b	$a+b$
Individual j				
	$-$	c	d	$c+d$
		$a+c$	$b+d$	p

($p=a+b+c+d$. The presence of a variable is denoted by $+$ and its absence by $-$.)

The rationale behind all of the coefficients given in Table A.1 is to count the total number of *relevant matches* between the two individuals on the p-variables. The number of coefficients is large because of the various possibilities for interpreting the phrase 'relevant matches'. Essentially the different interpretations arise because of two factors; the first and most important is the uncertainty over how to incorporate the number of negative matches (i.e. the term d) into the measures, and the second is the question of whether or not matches and mismatches should be equally weighted. Some measures exclude negative matches altogether (e.g. measure **3** in Table A.1), whilst others give higher weighting to matched pairs than to unmatched pairs (measures **4** and **5**). A number of the measures can be given a probabilistic interpretation; e.g. the simple matching coefficient, namely measure **1**, is simply the probability that a pair of individuals will achieve the same score on a randomly selected variable. Again measure **3** is the *conditional* probability that a pair of individuals will both have a randomly chosen variable present, given that 'absent–absent' matches are first discarded. Measures which give greater weight to matches than to mismatches generally have no such probabilistic interpretation. Perhaps the most widely used similarity measures are **1** and **3**. Fleiss (1975) shows that many similarity measures for present–absent data are equivalent to the kappa coefficient originally proposed as a measure of inter-observer agreement by Cohen (1960).

Distance Measures

Definition: a non-negative real value function, $d(x, y)$, of pairs of points of a set E is said to be a distance function (metric) for E if

1. $d(x, y) \geqslant 0 \; \forall \; x, y$
2. $d(x, y) = 0$ if and only if $x = y$
3. $d(x, y) = d(y, x)$
4. $d(x, y) \leqslant d(x, z) + d(y, z)$

The fourth of these conditions is the one which differentiates most between distance measures and similarity measures. This condition is referred to as the *metric inequality*, or more commonly as the *triangular inequality*

A list of some distance measures is given in Table A.2. Probably the most common of these, and the most familiar, is the Euclidean metric. Since this measure is badly affected by changing the scale of a variable, it is usually employed on the standardized data obtained by taking $z_{ik} = x_{ik}/\sigma_k$, where σ_k is the standard deviation of the k-th variable. The other distance measures are far less commonly used, although Carmichael and Sneath (1969) use the city block metric in their TAXMAP clustering procedure. The rationale given by these authors for this measure is that when, for example, two individuals are specified by two variables whose scale values are of equal value they should have the same distance apart whether (a) they are two units apart on each

Table A.2

Distance measures

1. Euclidean:

$$d_{ij} = \left[\sum_{k=1}^{p} (x_{ik} - x_{jk})^2 \right]^{1/2}$$

2. City block:

$$d_{ij} = \sum_{k=1}^{p} |x_{ik} - x_{jk}|$$

3. Sup. norm:

$$d_{ij}^{(\infty)} = \sup_{k=1\,\ldots\,p} \left\{ |x_{ik} - x_{jk}| \right\}$$

4. Minkowski:

$$d_{ij} = \left[\sum_{k=1}^{p} |x_{ik} - x_{jk}|^r \right]^{\frac{1}{r}}$$

5. Mahalanobis:

$$d_{ij} = (\mathbf{X}_i - \mathbf{X}_j)' \mathbf{W}^{-1} (\mathbf{X}_i - \mathbf{X}_j)$$

where $\mathbf{W}$ is the variance–covariance matrix and $\mathbf{X}_i'$ and $\mathbf{X}_j'$ are the $(1 \times p)$ vectors of scores for individuals i and j.

variable or (b) they are one unit apart on one variable and three on the other. The use of the city block metric meets this requirement. Mahalanobis distance has the useful property of being invariant under non-singular linear transformations of the data and so is unaffected by problems of scaling, etc.

Other 'distance' measures have been suggested which are not true metrics. For example, Lance and Williams (1966) suggest the measure

$$d_{ij} = \frac{\sum_{k=1}^{p} |x_{ik} - x_{jk}|}{\sum_{k=1}^{p} (x_{ik} + x_{jk})} \qquad \textbf{A.1}$$

and Duran and Odell (1974) list two other nonmetric measures, namely

$$d_{ij} = \left[\sum_{k=1}^{p} (\sqrt{x_{ik}} - \sqrt{x_{jk}})^2 \right]^{\frac{1}{2}} \qquad \textbf{A.2}$$

and

$$d_{ij} = \frac{1}{p} \left[\sum_{k=1}^{p} \frac{(x_{ik} - x_{jk})^2}{x_{ik} + x_{jk}} \right]^{\frac{1}{2}} \qquad \textbf{A.3}$$

The last of these measures, equation A.3, is generally known as the *coefficient of divergence*.

Inter-group Similarity and Distance Measures

The similarity and distance measures considered above were concerned with the measurement of these functions between pairs of *individuals*. In many cases the question arises as to how such measures may be made between two groups or clusters each containing several individuals. This may arise when the data consist *a priori* of groups and it is desired to examine in some way the relationships between the groups, or when the data have been subjected to some form of cluster analysis and the configuration of the resulting groups is to be investigated, say by using one of the ordination techniques described in Chapter 2.

Similarity (distance) between groups has been defined in many ways. One obvious procedure would be simply to take the 'average' similarity

102

or distance between pairs of individuals, one from each group. This would obviously only be useful if the concept of an average was acceptable for the particular inter-individual similarity or distance measure being used. A further intuitively reasonable procedure would be to measure inter-cluster similarity or distance using an inter-individual measure on the group means. For example, one might employ the Euclidean distance between the means as the measure for between groups.

Some other measures of inter-group distance are given below.

1. *Mahalanobis D^2*

$$d_{G_1 G_2} = (\bar{\mathbf{X}}_{G_1} - \bar{\mathbf{X}}_{G_2})' \mathbf{W}^{-1} (\bar{\mathbf{X}}_{G_1} - \bar{\mathbf{X}}_{G_2}) \qquad \textbf{A.4}$$

where $\mathbf{W}$ is the pooled within groups variance–covariance matrix, and $\bar{\mathbf{X}}'_{G_1}$ and $\bar{\mathbf{X}}'_{G_2}$ are the $(1 \times p)$ mean vectors of groups G_1 and G_2 respectively.

2. *Normal information radius* (Jardine and Sibson (1971))

$$d_{G_1 G_2} = \tfrac{1}{2} \log_2 \left\{ \frac{|\tfrac{1}{2}(\mathbf{W}_{G_1} + \mathbf{W}_{G_2})| + \tfrac{1}{4}(\bar{\mathbf{X}}_{G_1} - \bar{\mathbf{X}}_{G_2})'(\bar{\mathbf{X}}_{G_1} - \bar{\mathbf{X}}_{G_2})}{|\mathbf{W}_{G_1}|^{1/2}|\mathbf{W}_{G_2}|^{1/2}} \right\} \qquad \textbf{A.5}$$

where $\mathbf{W}_{G_1}$ and $\mathbf{W}_{G_2}$ are the variance–covariance matrices for groups G_1 and G_2 respectively, and $|\mathbf{W}_{G_i}|$ denotes the determinant of the matrix, $\mathbf{W}_{G_i}$ (When $\mathbf{W}_{G_1} = \mathbf{W}_{G_2}$ this reduces essentially to Mahalanobis distance.)

3. *Generalized distance for discrete variables*
For categorical data, Kurczynski (1970) defines the following generalized distance measure between groups,

$$d_{G_1 G_2} = (\mathbf{p}_{G_1} - \mathbf{p}_{G_2})' \mathbf{S}^{-1} (\mathbf{p}_{G_1} - \mathbf{p}_{G_2}) \qquad \textbf{A.6}$$

where $\mathbf{S}$ is the common sample variance–covariance matrix and $\mathbf{p}_{G_i}$ is a vector, the elements of which, $p_{G_i jk}$, give for Group G_i the proportions of individuals falling into category k of variable j. When the variables have a multinomial distribution this distance measure reduces to one given previously by Balakrishnan and Sanghwi (1968).

Appendix B: Computer Programs

Some of the methods discussed in this text are available in the many
'packages' of programs currently in use for multivariate analysis. Others
are not directly available but may be fairly easily implemented using
various published routines. In this appendix brief details are given of
where programs for many of the techniques are to be found; in addition
reference will be made to particular published algorithms which may
be of assistance in implementing those techniques not generally available.

Programs for the Methods discussed in Chapter 2

1. Principal components analysis
This method of analysis is available in all currently used packages for
multivariate analysis, e.g. BMD, SPSS, etc., although not all of these
allow some type of automatic plotting of principal component scores.

2. Principal co-ordinates analysis
This method of analysis is available in the GENSTAT package of
statistical programs produced at Rothamsted. This package contains
flexible routines for line printer plotting, enabling the principal co-
ordinate scores to be easily plotted. Information on this package is
available from the GENSTAT Secretary, Rothamsted Experimental
Station, Harpenden, Herts AL5 2JQ.

3. The biplot
The particular type of biplot described in Chapter 2 (page 23 is easily
implemented using one of the many efficient and widely available
routines for finding latent roots and vectors. The adaptation of most
principal component programs to perform such a plot would be a fairly
simple task.

4. Non-metric multidimensional scaling
Many programs are available for this technique. The most common
and the most widely used is the MDSCALE program developed by
Kruskal; this should be available at most computer installations.
  Guttman and Lingoes (*see* Lingoes (1972)) have produced a series of
programs which implement various adaptations of Kruskal's original

formulation of this method; these adaptations are known under the general heading of *smallest space analysis.*

A further program, INDSCAL, developed by Carrol (*see* Carroll (1973)) allows consideration of individual differences in judgements of similarity.

5. *Non-linear mapping*

An adaptation of the original program of Sammon for this method has been made by Howarth and is available from The Program Librarian, Imperial College Computer Centre, Exhibition Road, London SW7 23X.

A further program for this method which uses an alternative optimization procedure to that of Sammon is described in Chang and Lee (1973).

6. *Minimum spanning tree*

Ross (1969a) describes a routine which computes the minimum spanning tree of a distance matrix using the method of Prim, described by Gower and Ross (1969). Ross (1969b) also gives a program which prints the links of the minimum spanning tree in an order which is helpful in preparing it for display.

Programs for the Methods discussed in Chapter 3

1. *Agglomerative hierarchical techniques – dendrograms*

These techniques are available in several computer packages. The most widely used of these is the CLUSTAN suite of programs written by Dr D. Wishart, details of which may be obtained by writing to CLUSTAN, 16 Kingsburgh Road, Edinburgh, EH12 6DZ, Scotland.

An option available in this suite of programs which is especially useful is that of the automatic plotting of dendrograms using, for example, a microfilm plotter.

2. *The SHADE procedure*

A program implementing Ling's SHADE procedure for the direct graphical representation of similarity or distance matrices is available by writing to Dr R. F. Ling, Center for Mathematical Studies in Business and Economics, Graduate School of Business, University of Chicago, Chicago, Illinois 60637.

3. Cannonical variates analysis

This technique is implemented in many statistical packages for multivariate analysis, e.g. SPSS, and is also available as Computer Contribution 47 of the State Geological Survey, University of Kansas; this latter version was written by Reyment and Ramden.

4. Taxometric maps

This form of display was originally given by Carmichael and Sneath in association with their TAXMAP program for cluster analysis, and this program prints out the necessary information to produce a taxometric map. However, other clustering methods might easily be adapted so that they also gave this information.

Programs for the Methods discussed in Chapter 4

1. Probability plotting

The implementation of methods based on the technique of probability plotting is generally fairly simple. The major part of such a program will involve finding percentage points of various distributions. For example, for those techniques described in Chapter 4 (pages 65–81) we need the percentage points of the normal and gamma distributions. Routines which are useful in this context are described in Cunningham (1969), Bhattacharjee (1970), Sparks (1970), and Best and Roberts (1975).

Several routines which are useful for making gamma probability plots, including the estimation of necessary parameters, are available from Bell Telephone Laboratories. (See Gnanadesikan (1977).)

2. Andrews plots

The programming effort needed to implement this technique is very small, but an output device with relatively high precision, e.g. a microfilm plotter, is needed to draw the graphs. A listing of a FORTRAN program for this technique employing the microfilm software available at the University of London Computer Centre may be obtained by writing to the author of this text, Biometrics Unit, Institute of Psychiatry, Denmark Hill, London SE5.

3. Faces

The program written by Frith which produces cartoon faces on a microfilm plotter and again uses the microfilm software available at the University of London Computer Centre may also be obtained from the author at the address given above.

Line-printer Graph-plotting Routines

A large number of routines which enable graphs to be plotted on the line printer are available. The particular routine used for many of the line-printer diagrams reproduced in this text is due to Sparks (1971). The main difficulty in producing a reasonable line-printer plot is scale selection; some relevant routines in this context are those given by Rowlands and Pocock (1974) and Nelder (1976).

References

ALLEN, D. A. DE G. (1954) *Relaxation Methods* (New York: McGraw-Hill).

ANDERBERG, M. R. (1973) *Cluster Analysis for Applications* (New York: Academic Press).

ANDREWS, D. F. (1972) Plots of high dimensional data. *Biometrics*, **28**, 125–36.

ANDREWS, D. F., GNANADESIKAN, R. and WARNER, J. L. (1973) Methods for assessing multivariate normality. *Proc. International Symposium on Multivariate Analysis*, vol. 3, 95–116 (New York: Academic Press).

ANSCOMBE, F. J. (1960) Rejection of outliers. *Technometrics*, **2**, 123–46.

BALAKRISHNAN, V. and SANGHVI, L. D. (1968) Distance between populations on the basis of attribute data. *Biometrics*, **24**, 859–65.

BALL, G. H. and HALL, D. J. (1970) Some implications of interactive graphic computer systems for data analysis and statistics. *Technometrics*, **12**, 17–31.

BARNETT, V. (1975) Probability plotting methods and order statistics. *Appl. Statist.*, **24**, 95–107.

BEST, D. J. and ROBERTS, D. E. (1975) The percentage points of the chi-squared distribution. Algorithm AS 91. *Appl. Statist.*, **24**, 385–8.

BHATTACHARJEE, G. P. (1970) The incomplete gamma integral. Algorithm AS 32. *Appl. Statist.*, **19**, 285–7.

BLACKITH, R. E. and REYMENT, R. A. (1971) *Multivariate Morphometrics* (London: Academic Press).

BROCKINGTON, I. and LEFF, J. (1976) The incidence of schizo-affective psychosis. Submitted to *Archives of Psychiatry*.

CARMICHAEL, J. W., GEORGE, J. A. and JULIUS, R. S. (1968) Finding natural clusters. *Syst. Zool.*, **17**, 144–50.

CARMICHAEL, J. W. and SNEATH, P. N. A. (1969) Taxometric maps. *Syst. Zool.*, **18**, 402–15.

CARROLL, J. D. (1973) Individual differences and multidimensional scaling in *Multidimensional Scaling*, vol. 1, R. N. Shepard, A. K. Romney, and S. B. Nerlove (eds) (New York: Seminar Press).

CARROLL, J. D. and CHANG, J. J. (1970) Analysis of individual differences in multidimensional scaling via an N-way generalization of 'Eckart-Young' decomposition. *Psychometrika*, **35**, 283–319.

CATTELL, R. B. and COULTER, M. A. (1966) Principles of behavioural taxonomy and the mathematical basis of the taxonome computer program. *Brit. J. Math. Statist. Psychol.*, **19**, 237–69.

CHANG, C. L. and LEE, R. C. T. (1973) A heuristic relaxation method for non-linear mapping in cluster analysis. *IEEE Trans. on Systems, Man and Cybernetics*, **SMC-2**, 197–200.

CHERNOFF, H. (1973) Using faces to represent points in k-dimensional space graphically. *J. Am. Statist. Ass.*, **68,** 361–8.

COHEN, J. A. (1960) A coefficient of agreement for nominal scales. *Educ. and Psychol. Measurement*, **20,** 37–46.

COOMBS, C. H. (1964) *A Theory of Data* (New York: Wiley).

CORMACK, R. M. (1971) A review of classification. *J. Roy. Statist. Soc., Series A*, **134,** 321–67.

CUNNINGHAM, S. W. (1969) From normal integral to deviate. Algorithm AS 24. *Appl. Statist.*, **18,** 290–3.

DANIEL, C. (1959) Use of half-normal plots in interpreting factorial two-level experiments. *Technometrics*, **1,** 311–42.

DE LEY, J. (1962) Comparative biochemistry and enzymology in bacterial classification. *Sympos. Soc. Gen. Microbiol.*, **12,** 164–95.

DURAN, B. B. and ODELL, P. L. (1974) *Cluster Analysis: A Survey* (Heidelberg: Berlin; and New York: Springer-Verlag).

EVERITT, B. S. (1974) *Cluster Analysis* (London: Heinemann).

FISHER, R. A. (1936) The use of multiple measurements in taxonomic problems. *Annals of Eugenics*, **VII,** 179–84.

FLEISS, J. L. (1975) Measuring agreement between two judges on the presence or absence of a trait. *Biometrics*, **31,** 651–9.

FRASER, A. R. and KOVATS, M. (1966) Stereoscopic models of multivariate statistical data. *Biometrics*, **22,** 358–67.

FRITH, C. (1973) Personal communication.

GABRIEL, K. R. (1971) The biplot graphic display of matrices with applications to principal components analysis. *Biometrika*, **58,** 453–67.

GERSON, M. (1975) The techniques and uses of probability plotting. *The Statistician*, **4,** 235–57.

GNANADESIKAN, R. (1973) Graphical methods for informal inference in multivariate data analysis. Proceedings of the 39th session, *Bulletin of the International Statistical Institute*, 195–206.

(1977) *Methods for Statistical Data Analysis of Multivariate Observations* (New York: J. Wiley).

GNANADESIKAN, R. and KETTENRING, J. R. (1972) Robust estimates, residuals and outlier detection with multiresponse data. *Biometrics*, **28,** 81–124.

GNANADESIKAN, R. and LEE, E. T. (1970) Graphical techniques for internal comparisons amongst equal degree of freedom groupings in multiresponse experiments. *Biometrika*, **57,** 229–37.

GNANADESIKAN, R. and WILK, M. B. (1969) Data analytic methods in multivariate statistical analysis, in *Multivariate Analysis*, vol. II, P. R. Krishnaiah (ed.) (New York: Academic Press).

(1970) A probability plotting procedure for general analysis of variance. *Journal of Royal Statist. Soc. (Series B)*, **32,** 88–101.

GOODCHILD, N. A. and VIJAYAN, K. (1974) Significance tests in plots of multidimensional data in two dimensions. *Biometrics*, **30,** 209–10.

GOWER, J. C. (1966) Some distance properties of latent root and vector methods used in multivariate analysis. *Biometrika*, **53**, 325–38.

(1967) Multivariate analysis and multidimensional geometry. *The Statistician*, **17**, 13–25.

(1970) Classification and geology. *Res ISF*, **38**, 35–41.

(1971a) A general coefficient of similarity and some of its properties. *Biometrics*, **27**, 857–72.

(1971b) Statistical methods of comparing different multivariate analyses of the same data, in *Mathematics in the Archaeological and Historical Sciences*, F. R. Hodson, D. G. Kendall, and P. A. Tautu (eds) (Edinburgh: University Press).

GOWER, J. C. and BANFIELD, C. F. (1974) Goodness of fit criteria in cluster analysis and their empirical distributions. *The 8th International Biometric Conference*, Constanta, Romania.

GOWER, J. C. and ROSS, G. J. S. (1969) Minimum spanning trees and single linkage cluster analysis. *Appl. Statist.*, **18**, 54–64.

GREGSON, R. A. M. (1974) *Psychometrics of Similarity* (New York: Academic Press).

GUTTMAN, L. and LINGOES, J. C. (1967) Non-metric factor analysis: a rank reducing alternative to linear factor analysis. *Multiv. Behav. Res.*, **2**, 485–505.

HARTIGAN, J. A. (1967) Representation of similarity matrices by trees. *J. Amer. Statist. Assoc.*, **62**, 1140–58.

HAZEN, A. (1914) Storage to be provided in impounding reservoirs for municipal water supply. *Trans. Am. Soc. Civil Engrs*, **77**, 1539–659.

(1930) *Flood Flows. A Study of Frequencies and Magnitudes* (New York: Columbia University Press).

HEALEY, J. R. (1968) Multivariate normal plotting. *Appl. Statist.*, **17**, 157–61.

HILLS, M. (1969) On looking at large correlation matrices. *Biometrika*, **56**, 149–53.

HOWARTH, R. J. (1973) Preliminary assessment of a non-linear mapping algorithm in a geological context. *Mathematical Geology*, **5**, 39–57.

HUFF, D. (1954) *How to Lie with Statistics* (New York: Norton).

JARDINE, C. J., JARDINE, N. and SIBSON, R. (1967) The structure and construction of taxonomic hierarchies. *Math. Biosc.*, **1**, 173–9.

JARDINE, N. and SIBSON, R. (1968) The construction of hierarchic and non-hierarchic classifications. *Com J.*, **11**, 117–84.

(1971) *Mathematical Taxonomy* (New York: Wiley).

JASPER, R. (1949) Unpublished technical report.

KENDALL, M. G. (1975) *Multivariate Analysis* (London: Griffin).

KRUSKAL, J. B. (1964a) Multidimensional scaling by optimizing goodness of fit to non-metric hypotheses. *Psychometrika*, **29**, 1–27.

(1964b) Non-metric multidimensional scaling: a numerical method. *Psychometrika*, **29**, 115–29.

KRUSKAL, J. B. and CARROL, J. D. (1969) Geometrical models and badness-of-fit functions, in *Multivariate Analysis*, vol. II, P. R. Krishnaiah (ed.) (New York: Academic Press).

KRZANOWSKI, W. J. (1971) A comparison of some distance measures applicable to multinomial data using a rotational fit technique. *Biometrics*, **27**, 1062–8.

KURCZYNSKI, T. W. (1970) Generalized distance and discrete variables. *Biometrics*, **26**, 525–34.

LANCE, G. N. and WILLIAMS, W. T. (1966) Computer programs for hierarchial polythetic classification. *Comp. J.*, **9**, 60–4.

LAWLEY, D. N. and MAXWELL, A. E. (1971) *Factor Analysis as a Statistical Method* (London: Butterworths).

LING, R. F. (1973) A computer generated aid for cluster analysis. *Communications of the ACM*, **16**, 355–61.

LINGOES, J. C. (1972) A general survey of the Guttman–Lingoes non-metric program series, in *Multidimensional Scaling*, vol. 1, R. N. Shepard, A. K. Romney, and S. B. Nerlove (eds) (New York: Seminar Press).

LOBERMAN, H. and WEINBERGER, A. (1957) Formal procedures for connecting terminals with a minimum total wire length. *J. Assoc. Comp. Mach.*, **4**, 428–37.

MAHALANOBIS, P. C., MAJUMBAR, D. N. and RAC, C. R. (1949) Anthropometric survey of the United Provinces. *Sankhya*, **9**, 89–324.

MARDIA, K. V. (1975) Assessment of multinormality and the robustness of Hotelling's test. *Appl. Statist.*, **24**, 163–71.

MAXWELL, A. E. (1961) Cannonical variate analysis when the variables are dichotomous. *Educ. and Psychol. Measure.*, **21**, 259–71.

MOSS, W. W. (1967) Some new analytic and graphic approaches to numerical taxonomy, with an example from the Dermanyssidae (Acari). *Syst. Zool.*, **16**, 177–207.

NATHANSON, J. A. (1971) Applications of multivariate analysis in astronomy. *Appl. Statist.*, **20**, 239–49.

NELDER, J. A. (1976) A simple algorithm for scaling graphs. Algorithm AS 96. *Appl. Statist.*, **25**, 94–6.

PEARSON, E. S. (1956) Some aspects of the geometry of statistics. The use of visual presentation in understanding the theory and application of mathematical statistics. *J. Roy. Statist. Soc., Series A*, **119**, 5–146.

PRESS, S. J. (1972) *Applied Multivariate Analysis* (New York: Holt and Reinhart).

PRIM, R. C. (1957) Shortest connection matrix network and some generalizations. *Bell System Tech. J.*, **36**, 1389–401.

ROHLF, F. J. (1970) Adaptive hierarchial clustering schemes. *Syst. Zool.*, **19**, 58–82.

ROSS, G. J. S. (1969a) Minimum spanning trees. Algorithm AS 13. *Appl. Statist.*, **18**, 103–4.

(1969b) Printing the minimum spanning tree. Algorithm AS 14. *Appl. Statist.*, **18**, 105–6.

ROWLANDS, G. J. and POCOCK, R. M. (1974) Remarks on scale selection for scatter diagrams. ASR 10. *Appl. Statist.*, **23**, 248–9.

SAMMON, J. W. (1969) A non-linear mapping for data structure analysis. *IEEE Trans. Computers*, **C18**, 401–9.

SHEPARD, R. N. (1962a) The analysis of proximities: multidimensional scaling with an unknown distance function I. *Psychometrika*, **27**, 219–46.

(1962b) The analysis of proximities: multidimensional scaling with an unknown distance function II. *Psychometrika*, **27**, 125–39.

(1973) Introduction to vol. 1, in *Multidimensional Scaling*, vol. 1, R. N. Shepard, A. K. Romney, and S. B. Nerlove (eds) (New York: Seminar Press).

(1974) Representation of structure in similarity data; problems and prospects. *Psychometrika*, **39**, 373–421.

SNEATH, P. H. A. and SOKAL, R. R. (1973) *Numerical Taxonomy* (San Francisco: W. H. Freeman).

SOKAL, R. R. and RHOLF, F. J. (1962) The comparison of dendrograms by objective methods. *Taxon*, **11**, 33–40.

SPARKS, D. N. (1970) Half-normal plotting. Algorithm AS 30. *Appl. Statist.*, **19**, 192–6.

(1971) Scatter diagram plotting. Algorithm AS 44. *Appl. Statist.*, **20**, 327–31.

TORGERSON, W. S. (1952) Multidimensional scaling 1. Theory and method. *Psychometrika*, **17**, 401–19.

TUCKER, L. R. (1972) Relations between multidimensional scaling and three-mode factor analysis. *Psychometrika*, **37**, 3–27.

TUKEY, J. W. (1970) *Exploratory Data Analysis* (limited preliminary edition) (Reading: Addison Wesley).

WILK, M. B. and GNANADESIKAN, R. (1961) Graphical analysis of multi-response experimental data using ordered distances. *Prac. Nat. Acad. Sci. U.S.A.*, **47**, 1209–12.

(1964) Graphical methods for internal comparisons in multi-response experiments. *Ann. Math. Statist.*, **35**, 613–31.

(1968) Probability plotting methods for the analysis of data. *Biometrika*, **55**, 1–17.

WILK, M. B., GNANADESIKAN, R. and HUYETT, M. J. (1962a) Probability plots for gamma distribution. *Technometrics*, **4**, 1–20.

(1962b) Estimation of the parameters of gamma distribution using order statistics. *Biometrika*, **49**, 525–45.

WOLFE, J. H. (1970) Pattern clustering by multivariate mixture analysis. *Multiv. Behav. Res.*, **5**, 329–50.

Index